Behaviour and Social Evolution of Wasps

The Communal Aggregation Hypothesis

YOSIAKI ITÔ

Laboratory of Applied Entomology and Nematology
Nagoya University, Japan

D0103300

Oxford   New York   Tokyo
OXFORD UNIVERSITY PRESS
1993

Oxford University Press, Walton Street, Oxford OX2 6DP

Oxford New York Toronto
Delhi Bombay Calcutta Madras Karachi
Kuala Lumpur Singapore Hong Kong Tokyo
Nairobi Dar es Salaam Cape Town
Melbourne Auckland Madrid
and associated companies in
Berlin Ibadan

Oxford is a trade mark of Oxford University Press

Published in the United States
by Oxford University Press Inc., New York

A catalogue record for this book is available from the British Library

Library of Congress Cataloging in Publication Data
Itô, Yosiaki, 1930–
Behaviour and social evolution of wasps: the communal aggregation hypothesis/Yosiaki Itô.
(Oxford series in ecology and evolution)
Includes bibliographical references and index.
1. Wasps—behaviour. 2. Insect societies. 3. Behaviour evolution.
4. Kin selection (Evolution) I. Title. II. Series.
QL568.V5I77 1993 595.79'8—dc20 92-25782
ISBN 0–19–854683–1 (hbk.)
ISBN 0–19–854046–9 (pbk.)

Typeset by Cotswold Typesetting Limited, Gloucester
Printed and bound in Great Britain by
Biddles Ltd, Guildford and King's Lynn

Acknowledgements

Ever since I began work in entomology at the National Institute of Agricultural Sciences, Tokyo, in 1950, I have concentrated on the population ecology of insects. However, my interest in the evolution of insect societies was first aroused by the Japanese book *Ningen-izen no syakai* (*Infrahuman societies*) (Imanishi 1951). During the period 1950–1980, I followed the development of studies on social wasps through frequent discussions with the pioneers in research on social insects in Japan: Kunio Iwata, Katsuji Tsuneki, Kimio Yoshikawa, Rokuya Morimoto, and Syoiti F. Sakagami. In 1980 I began to study paper wasps, *Polistes* spp., with my students Eiichi Kasuya and Yoshinori Hibino.

In 1982, I commenced studies on the Okinawan paper wasp, *Ropalidia fasciata*, the discussion of which is an important part of this book. I thank my collaborators, Sôichi Yamane, Seiki Yamane, and Osamu Iwahashi, for permitting me to include many results from our joint studies.

The Smithsonian Tropical Research Institute gave me the opportunity to spend two months in Panama. Without this I could not have developed my ideas on the evolution of wasp sociality in the wet tropics. I thank my old friend, Henk Wolda, for making this possible.

My frequent visits to Australia were made possible by the Fellowship of Australian Vice Chancellors' Committee, and grants for overseas scientific surveys (numbers 61042002 and 62041023) from the Japan Ministry of Education, Science, and Culture. For the former I thank Roger Kitching and Jiro Kikkawa, who recommended me to the Committee.

My work was also funded by a Grant-in-Aid for Special Project Research on Biological Aspects of Optimal Strategy and Social Structure (1983–1986; Head Ei Teramoto), and for Scientific Research (number 62304002) from the Japan Ministry of Education, Science, and Culture. Using the former fund, we were able to sponsor an international symposium in Kyoto in 1986, at which I had an opportunity to discuss the problems covered in this book with a number of most active theorists: R. H. Crozier, Raghavendra Gadagkar, W. D. Hamilton, and M. J. West-Eberhard.

The identification of wasps collected in Panama and Australia was made by the late O. W. Richards, Seiki Yamane, Jun'ichi Kojima, and J. M. Carpenter.

In writing and revising the manuscript, I have been helped by many friends. Critical reading of my early notes by S. F. Sakagami, Eiichi Kasuya, Mary Jane West-Eberhard, and C. K. Starr, and many discussions with Naomi

Pierce were most helpful. Sôichi Yamane, Makoto Matsuura, Jun'ichi Kojima, and Tadashi Suzuki provided me with a lot of the information on the biology of vespid wasps.

I thank J. P. Spradbery for reading an early draft of the manuscript and making many scientific comments and text corrections. Thanks are also due to Robert May, Paul Harvey, and two anonymous referees for their help in improving the manuscript, and to the staff of the Oxford University Press for their kind arrangements.

I thank Tôkai Daigaku Syuppankai (Tôkai University Press), Tokyo, who kindly permitted the reproduction of many parts of the text, tables, and figures from *Karibati no syakai-sinka*, which gave rise to this book.

Last, but not least, I thank my wife, Ayako, for giving me the freedom to visit many places on many occasions for field studies.

Nagoya   Y.I.
March 1992

Contents

1

Introduction

This book is a development of the ideas that I first presented in *Karibati no syakai-sinka* (*Social evolution of wasps*, published by Tôkai University Press, Tokyo, in 1986).

The social insects represent the peak of social integration among invertebrate animals. Thus honey-bee queens cannot live alone, but must always be helped by workers. The workers do not mate, though they can lay male-destined (haploid) eggs when their queen is lost. This is however, exceptional; the workers usually devote their lives to feeding their queen's young and defending their nest, sometimes sacrificing their own lives in the process. Such social systems, characterized by the existence of complete or incomplete division of reproductive labour (the coexistence of two or more 'castes') and by co-operation among colony members, are defined as 'eusocial' (Wilson 1975).

The wasp family Vespidae is particularly suitable for studying the evolution of eusociality, because it contains examples of many different stages in the evolution of social behaviour: infra-eusocial species, primitively eusocial species (in which there is no morphological difference between queen and worker), and fully developed eusocial species (in which the queens are morphologically distinguishable from workers).

This book is devoted to the social lives of the Vespidae. It is not however, a textbook on eusocial wasps nor a comprehensive review of the literature. Rather, the book aims to present the author's opinions about the evolution and maintenance of eusociality in the Vespidae, based mainly on results of field studies in tropical and subtropical areas.

My aim is to emphasize two points: first, that the coexistence of two or more egg-layers (multiple 'queens') in social insect colonies is much more common, especially in the tropics, than is currently thought; and secondly, that among three major hypotheses about the evolution of eusociality, namely kin-selection, parental manipulation, and mutualism (Chapter 2), the third—mutualism—plays a larger role than current literature suggests.

My ideas have particular bearing on a controversy over two hypotheses about how multi-queen social systems may have evolved in the eusocial wasps (Chapter 9). West-Eberhard (1978*b*) proposed the *polygynous family hypothesis*, which states that 'it is possible that the highly specialized, worker-containing, polygynous colonies of some Polibiini could have developed directly from polygynous, primitively social groups without passing through a monogynous stage'. Conversely, Carpenter (1989) considered that 'there is no evidence for a polygynous transition from the caste-containing group without an intervening monogynous stage.' A principal aim of this book is to suggest that much more evidence to support West-Eberhard's view may be found as detailed studies on social wasps in the wet tropics become more numerous.

Hamilton's (1964) kin-selection hypothesis undoubtedly created a new era in thinking about social evolution. However, the role of kin-selection has sometimes been over-emphasized in discussions of insect eusociality, especially during the 1970s, despite a comprehensive review by Hamilton himself (1972). Although many authors (e.g. West-Eberhard 1978*b*; Andersson 1989; Krebs and Davies 1987) have now recognized that the various hypotheses are not mutually exclusive, but rather that they all may have contributed to the evolution of eusociality, many researchers still seem to be concentrating their efforts on the role of dominance hierarchies. This mechanism limits the number of reproductives in the colony and thereby keeps intracolony relatedness high, a prerequisite for the operation of kin selection.

However, in many tropical polistine wasps, some vespine wasps, and many ants, there are species which have multi-queen colonies. Intracolony relatedness is likely to be low for these species. Thus more attention must be paid to mutualism (*sensu* Lin and Michener 1972).

In this book I have made no attempt to present an exhaustive review of all the literature that is relevant to such theoretical debates. Instead, I have selectively focused on studies dealing with the relation between mutualism and the evolution of social behaviour. Of course there are many other papers on related topics, including works in which dominance hierarchies, limitation of oviposition to a single queen, and/or kin-selection are stressed, but readers will easily be able to find this literature from review papers cited in this book (e.g. Wilson 1971, 1975; Hamilton 1972; Brockmann 1984; Andersson 1984; Gadagkar 1990; Ross and Matthews 1991). This book is a deliberate attempt to swing the pendulum the other way. My evidence is of course incomplete; although my observations cover many tropical species for which no data were available previously, the studies were short and intermittent. But if the viewpoint expressed here can stimulate new approaches to future field and theoretical research, it will be worthwhile, even if the findings do not support my hypothesis.

Chapter 2 introduces, for the readers who are unfamiliar with these insects, the systematics of the major taxa of interest, and surveys two hypothetical routes for the evolution of eusociality in the Vespidae. It also explains the terminology used in the book to indicate different types of colony foundation and social structure.

In Chapters 3 and 4, the major hypotheses for the evolution of eusociality and some problems they raise are introduced. These chapters provide the theoretical background for Chapters 6–13. Chapter 5, summarizes data on the frequencies of intranidal dominance-aggressive acts, percentage of pleometrotic colonies, and survival rates of pleometrotic colonies compared with haplometrotic ones in many species of the Polistinae. The terms pleometrotic and haplometrotic refer to colonies which have multiple and single egg-layers respectively (see Fig. 2.4 for more detailed definitions). Most of the data were collected during the research described in Chapters 6, 7, 8, and 9, but I believe it is better to present the summary chapter before chapters on each species, because this summary may simplify the interspecific comparisons.

Chapter 6 introduces the results of my studies on the Okinawan paper wasp, *Ropalidia fasciata*. This species appears to represent an intermediate position between haplometrotic *Polistes* and more pleometrotic, tropical *Ropalidia*. Chapters 7 and 8 describe research on the social biology of wasps in Panama and Australia, many of the results support the view that pleometrosis is common in the tropics.

Chapter 9 examines swarm-founding tropical Polistinae, while Chapter 10 deals with pleometrosis in the Vespinae, Stenogastrinae, and *Belonogaster* (Polistinae).

In Chapter 11, I discuss the conditions which have favoured the evolution of pleometrosis using an inclusive fitness model, and I also consider pleometrosis in bees, ants, and termites.

In Chapters 12 and 13, I explore the hypothesis of the manipulation of progeny by foundress groups which is one of the key elements in my thinking. I also discuss ways in which multi-queen colony systems could have evolved through facultative pleometrosis.

In short, the principal aim of this book is to stress the role of mutualism in the evolution of insect eusociality, and to suggest that greater recognition of this role will lead to a better evolutionary understanding of the multi-queen social systems seen in tropical wasps. This social system is still not satisfactorily explained in the literature of evolutionary ecology and is likely to remain a subject for debate in insect behavioural ecology for some time to come.

2

Systematics and sociality of wasps

2.1 The Hymenoptera

The order Hymenoptera is divided into two suborders, Symphyta and Apocrita, and the latter is further divided into Terebrantia and Aculeata. Eusociality only occurs in the Aculeata and one group of the Terebrantia, the Chalcidoidea, in which one species, *Capidosomopsis tanytmemus* has sterile, defensive larvae, like aphid soldiers (Cruz 1981).

The designations 'bee' and 'wasp' respectively refer to phytophagous and mainly carnivorous[1] groups of the Aculeata. The wasps include species belonging to many different systematic groups (e.g. Bethyloidea, Sphecormis group, and Vespoidea; in addition, species of Symphyta and Terebrantia are also called wasps), but all the bee species belong to a single taxon, the Apiformes (Fig. 2.1).

Figure 2.1 indicates that eusociality has evolved in at least six different taxa. In the Sphecoidea, the bee group (Apiformes) is thought to have evolved eusociality at least eight times (in Halictidae, Anthophoridae, and Apidae), while in the Sphecidae this occurred only once, in the genus *Microstigmus*, which shows signs of primitive eusociality (Matthews 1968). Conversely, in the Formicidae, which contains 12 000–14 000 species (Wilson 1971), almost all living species exhibit developed eusociality in which queens and workers are morphologically distinguishable (the exceptions are those species which lack queens, e.g. *Pristomyrmex pungens* (Tsuji 1988), and social parasites which lack workers, e.g. *Teleutomyrmex schneideri* (Wilson 1975)).

2.2 The Vespidae

The possible phylogenic relationships of the subfamilies of the Vespidae (formerly Vespoidae) are shown in Fig. 2.2 (Carpenter 1982).

---

[1] Carnivorous refers here to larvae that feed on food derived from animals. The adults of most carnivorous wasps feed on honey.

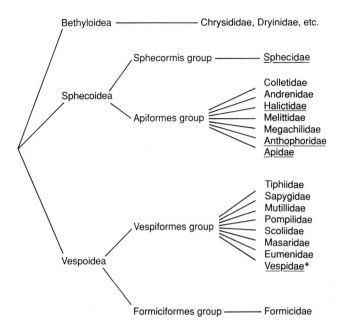

Bethyloidea ──────────── Chrysididae, Dryinidae, etc.

Fig. 2.1 Phylogeny of the Aculeata, based on Brothers (1974). Some families and subfamilies have been omitted (see also Snelling 1981). Taxa which include eusocial species are underlined.
*For a new classification of the Vespidae, see Fig. 2.2.

The subfamily Eumeninae has some subsocial species (exhibiting progressive provisioning and contact of mother and larvae) and many solitary species. Although eusociality is not known in this subgenus, there is a possibility that rudimentary division of reproductive labour may occur in some tropical genera. The neotropical species *Zethus miniatus*, which is basically solitary, sometimes exhibits group nesting (West-Eberhard 1987).

The subfamily Stenogastrinae (Fig. 2.2), which consists of 67 species (Akre 1982), is an interesting group. Although a morphologically distinguishable worker caste has not been found, several female adults belonging to the same generation or two successive generations often coexist in a nest, one of them being the principal egg-layer. Stenogastrine eusociality is not only rudimentary, but may also show special adaptations to the tropical rainforest environment (see Chapter 10).

The subfamily Polistinae comprises more than 380 known species, all of which (except a few social parasites) are eusocial. The Polistinae were divided into three tribes by Richards (1978a); the Polybiini, Polistini, and Ropalidiini. But Carpenter (1989) abandoned this division. Richards' Polistini (one genus) and Ropalidiini (one genus) may certainly be independent groups, but his

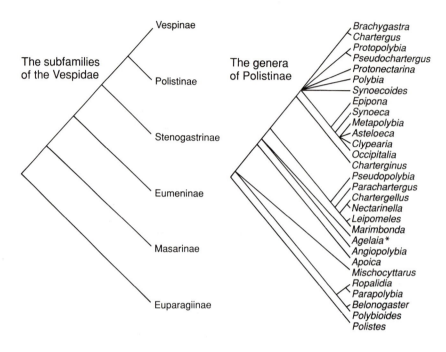

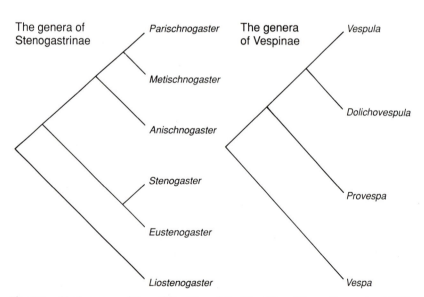

Fig. 2.2 Cladograms of the subfamilies of the Vespidae. (From Carpenter 1989). *The old genus name *Stelopolybia* is used in the text.

Polybiini (25 genera) may contain polyphyletic groups. For example, a large neotropical genus *Mischocyttarus* seems to be distinctly different from other polybiine genera (Fig. 2.2; see also Carpenter 1991). However, many neotropical genera of the Polybiini may belong to a single group (except for *Mischocyttarus*), for example *Metapolybia, Polybia, Protopolybia, Brachygastra*, and *Stelopolybia* (Carpenter's *Agelaia*)—which make enveloped nests like hornets. For convenience the term 'polybiine wasps' is used in this book to describe this group (it is possible that most of the neotropical genera above *Angiopolybia* in Fig. 2.2 might be included in this group when their social lives become known).

The genus *Ropalidia* is a large group containing about 126 species (Akre 1982) (Chapters 6, 8, and 9). This genus includes *independent-founding* species (in which new nests are founded by 'foundresses' which become queens after the emergence of progeny, as in *Polistes*) and *swarm-founding* species (in which nests are founded by groups of queen(s) and workers). The independent-founding species include two types of nest foundation: foundation by a single foundress (*single-female-founding*) and foundation by a group of foundresses (*multi-female-founding*).

The genus *Polistes* contains 203 eusocial species and three social parasites (Akre 1982). All known eusocial species found their nests by independent-founding. In the temperate zone, only inseminated females overwinter and found nests the next spring. Their first progeny are, as a rule, females, and these become workers. Most temperate zone *Polistes* found nests by single-female-founding, but many tropical species are multi-female-founding (Chapters 7 and 8).

Parapolybia is a small Asian genus (six species). *Parapolybia indica* (Sekijima *et al.* 1980) and possibly *P. varia* (S. Yamane, personal communication) found nests in mainland Japan by single-female-founding. In Taiwan, *P. varia*'s nests are established by foundress associations.

All known species of the African genus *Belonogaster* (approximately 35 species) and the large American genus *Mischocyttarus* (202 species) are independent-founding and most of them may be multi-female-founding (Chapters 7 and 10).

The 'polybiine wasps' (approximately 200 species) are unique among the Vespidae, because most of them reproduce by swarming, and their colonies often exceed several thousand adults. Another unique character of this group is that most of the colonies for which colony composition is known have, at least in some stage of their colony cycle, multiple queens (Chapter 9).

The Vespinae have reached the most developed level of eusociality among the Vespidae, because in most species queens are clearly distinguishable from workers. Although, thus far, all of them have been considered to be single-female, independent-founding species, recent observations strongly indicate the need for some change in this view (Chapter 10).

2.3 Terminology

The colony cycle of temperate *Polistes* or Vespinae has long been considered to be representative of eusocial wasps. Their typical colony cycle is explained below (Fig. 2.3).

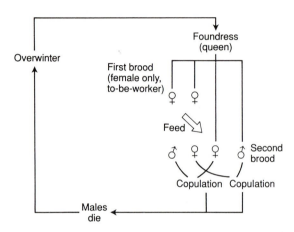

Fig. 2.3 A 'typical' colony cycle for temperate zone eusocial wasps. Overwintered females found nests alone or in groups in the spring. All of their first-brood progeny are females and these assist their mothers as 'workers' (arrow). In late summer, males and females, which are destined to be the next year's foundresses, emerge. After copulation, only to-be-foundress females overwinter.

In the typical colony cycle of these species, only the females which emerge in autumn will mate and overwinter. In spring, the overwintered females build new nests alone or in groups. These females are called '*foundresses*'. The first group of progeny produced by the foundresses (called 'the first brood' hereafter) is, as a rule, all female. These females remain on natal nests and become workers. The foundresses thereafter cease their extranidal activities (activities outside the nest) and, as queens, devote themselves to oviposition. In mid-summer, the males and second-brood females emerge. These second-brood females mate with the males and overwinter. The classification of nest-founding is shown in Fig. 2.4. With independent-founding, examples in which a single foundress establishes her nest and feeds her first-brood larvae, are called single-female-founding.

Multi-female-founding includes two types: (1) a single female initiates a nest, and then several females join it (as in *Polistes fuscatus*; West-Eberhard 1969); and (2) each nest is founded by a group of foundresses from the initial stage; these foundresses usually having emerged from the same nest and overwintered in a group (as in *Ropalidia fasciata*, see Chapter 6). The latter

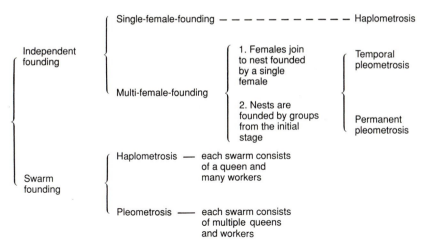

Fig. 2.4   Major types of nest founding and colony composition. Haplometrosis means that each colony has a single egg-layer (queen), while pleometrosis means that at least some colonies have multiple egg-layers. Many social insect researchers (e.g. Hölldobler and Wilson 1977) have used different terminology, in which single-female-founding and multi-female-founding were called haplometrosis and pleometrosis, respectively, and the existence of multiple egg-layers was called polygyny. But, as stated by Richards and Richards (1951), the terms polygyny and monogyny should be used to distinguish mating systems. Richards and Richards' (1951) terminology is used here, that is, pleometrosis means 'presence of more than one egg-laying, fertilized female in the same nest', and haplometrosis means 'presence of only one such female'.

type may represent an intermediate position between independent-founding and swarm-founding. For both these cases, pleometrosis may be temporal or permanent. In the former, foundresses other than a single, dominant foundress may disappear or become functional workers. In the latter, however, multiple egg-layers can coexist even after the emergence of first-brood females. If multiple foundresses continue to coexist but oviposition is monopolized by a single female, it is termed 'functional haplometrosis'.

Swarm-founding includes nest foundation by single queen/multiple worker swarms (as in the honey-bee) and also by multiple queen/multiple worker swarms.

2.4   Hypothetical routes towards eusociality in the Hymenoptera

Subsocial route

There are two main aspects of evolution of eusociality in the Hymenoptera that need to be distinguished. The first concerns processes: the attempt to trace, regardless of the causal relations, the possible route of evolution of eusociality. This is discussed below. The second concerns causality: how the altruistic traits of workers, which do not have any progeny of their own, but devote their lives to tending their nestmates, can evolve.

An evolutionary route to eusociality in the Hymenoptera was first presented by W. M. Wheeler (1923) and this classic scheme is still useful (Fig. 2.5).

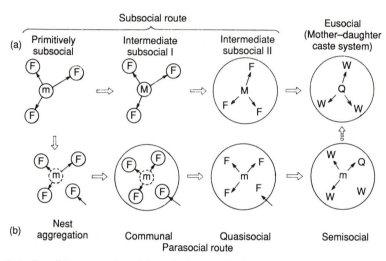

Fig. 2.5 Possible route of social evolution in the Hymenoptera, presented by (a) Wheeler (1923) and (b) Michener (1958). M, mother (adult; m indicates death of the mother); F, daughter (adult); Q, queen; W, worker. Large circles indicate nests made by the mother. Small circles indicate mother's nest; dashed ones are those within nest aggregation. Arrows from outside the large circles indicate female adults that have joined the nest. (Based on Sakagami (1970), using Wilson's (1971) terminology).

1. Among solitary parasitic wasps, (most of which find a host, then lay eggs on (or in) the host body and leave), there are some species which provide food for their progeny. At this stage (termed the mass-provisioning stage), the mother wasp builds a nest, stocks it with food, lays an egg on or near the food, and closes the nest. Although in the mass-provisioning stage there is no social contact between the two successive generations, a simple form of maternal care has evolved.

2. Among mass-provisioning species, some exhibit further development in the evolution of maternal care—progressive-provisioning. Here, a mother lays her egg(s) before or during the collection of food and continues to provide fresh food for her larva(e). The mother makes contact with her immature progeny, but she may leave the nest when the larvae pupate. Social contact

between the adults of the two successive generations does not occur. Wheeler called this the 'subsocial stage'.

3. Among progressive-provisioning species, coexistence of the adults of two successive generations, mother and daughters, occurs. In this case, the daughters perform most of the foraging and nest building activities, while the mother devotes herself to oviposition after the emergence of daughters. This is the division of reproductive labour between two successive generations and represents true eusociality (mother–daughter caste system).

Thus, in the subsocial route, progressive increase in maternal care has resulted in the coexistence of mother and progeny, and this has provided a condition for the evolution of castes. Here eusociality is considered to be a result of the development of family systems. Evidence for the evolution of eusociality through the subsocial route has mainly been obtained from wasps, the Sphecidae and Vespidae.

A possible route of transition from solitary to subsocial lives in wasps and bees was proposed by Iwata (1942) and later modified by Evans (1953). The transition from mother–daughter coexistence (intermediate subsocial II of Fig. 2.5) to eusociality was demonstrated by Iwata (1938) and Skaife (1953) in two allodapine bees, *Braunsapis sauteriella* and *Allodape angulata*, respectively. These are progressive-provisioning bees, and are rather exceptional because most bees are mass-provisioning species.

Parasocial route

In 1956 (published as a paper in 1958), Michener presented another possible route, now called the 'parasocial route', for the evolution of eusociality. The parasocial route is thought to have proceeded as follows:

1. In some apiform species (bees) it has been well known that several hundred female bees dig their nests in aggregations at a particular site, primarily because of shortages of suitable nesting sites (for nest aggregation of sphecoid wasps, see Tsuneki 1965). In many species each female digs a separate nest hole, but in some species several females use a common entrance and each female occupies her own gallery for provisioning her progeny. This situation is called the 'communal stage'.

2. Some communal species have appeared in which each female does not exclusively use her own gallery but provides food to the larvae in every gallery (e.g. *Halictus tumulorum*, Sakagami 1970). Michener called this step the 'quasisocial stage'.

3. A dominance hierarchy and the rudimentary reproductive division of labour among cofoundresses then appeared, possibly based on physiological variations arising from different nutritional conditions during growth.

Here, physiologically weaker females perform more extranidal tasks, while stronger females lay most of the eggs. This within-generation caste system is called the 'semisocial stage'.

4. When the daughters which emerged in such semisocial colonies became workers, due to parental manipulation, eusociality had developed (see the broken arrow connecting D′ and D in Fig. 2.5).

In my book *Comparative ecology* (Itô 1980; Chapter 5), the validity of the parasocial route in the Vespidae is questioned, because of the apparent absence of an explanation for the transition from the semisocial to the eusocial stage. However, recent information which suggests the existence of quasisocial and semisocial species in Eumeninae (*Zethus* etc., West-Eberhard 1987; *Auplopus semialatus*, Wcislo *et al.* 1988) strongly supports the possibility of a parasocial route in the Vespidae. One aim of the present book is to put forward an hypothesis which attempts to explain the transition from semisocial to eusocial organization in insects (Chapter 12).

3

Theories on the evolution of eusociality

3.1   Charles Darwin's dilemma

In *The origin of species*, Darwin wrote of 'one special difficulty, which at first appeared to me insuperable, and actually fatal to my whole theory.' He was referring to the existence of sterile workers among the social insects. As the worker is 'absolutely sterile', 'it could never have transmitted successively acquired modifications of structure or instinct to its progeny.' For the evolution of a new varied trait to occur, individuals who bear this trait must leave more progeny than others; but the workers leave no offspring. To rescue his theory, Darwin introduced the idea of natural selection applied to the family, rather than to the individual. His logic seems to be correct, even to the present day: 'breeders of cattle wish the flesh and fat to be well marbled together; the animal has been slaughtered, but the breeder goes with confidence to the same family' (Darwin 1859; p. 258). Thus, Darwin clearly noticed the successful reproduction of kin which, by common descent, shares genes identical to those of the individual which leaves no offspring. Darwin also wrote that 'by the long-continued selection of the fertile parents which produced most neuters with the profitable modification, all the neuters ultimately come to have the desired character'. He thus approached the idea of parental manipulation presented by Alexander (1974).

However, during the 100 years after Darwin's time most ecologists were restricted to a pre-Darwinian viewpoint. Without any logical explanation, they considered that 'even a trait which is unsuitable for the bearer can evolve if the trait is beneficial for the whole society.' While a few theorists (Fisher 1930; Haldane 1932; Sturtevant 1938; Wright 1945) had approached the correct answer before the end of World War II, it was not until 1964 that Hamilton developed the final solution to this problem.

3.2 Hamilton's theory of inclusive fitness and kin-selection

Fitness, which is the criterion determining whether or not a trait evolves, was traditionally measured by the number of offspring that were produced by an individual and reached the reproductive stage. However, Hamilton (1964) proposed that true fitness should be measured by the number of copies of a gene found in the progeny generation.

Thus, even if an individual (A) leaves no offspring due to its altruism, if B, A's kin, which shares the gene in question with A by common descent, leaves many offspring, the number of copies of the gene in the progeny generation increases. Thus, true fitness must be defined as fitness in the absence of any social interaction, minus the cost to fitness of the altruism, plus the increment in the fitness of the recipient, multiplied by relatedness. Hamilton called this *inclusive fitness*. Maynard Smith named selection by this route, ('bypath') *kin selection*.

Inclusive fitness (I) can be represented by

$$I = W_{0(A)} - \Delta W_A + \Sigma \Delta W_i r_i \qquad (3.1)$$

where $W_{0(A)}$ is the fitness of an individual A in the absence of any social interaction (i.e. A does not help and is not helped by any other individuals), ΔW_A is the fractional decrease in the fitness of A due to its altruism, $\Sigma \Delta W_i$ is the fractional increase in the fitness of the ith individual ($i \neq A$) due to the altruistic act, and r_i is the coefficient of relatedness (i.e. the probability that A and the ith individual will share a copy of a particular gene identical by descent). This is therefore a quantitative model of the evolution of altruism among animals, based on the concept of inclusive fitness.

If the cost of an altruistic act, ΔW_A, is C, and the benefit ΔW (supposing there is only one recipient) is B, then the gene causing the donor to act altruistically will increase if $B/C > 1/r$ or $Br - C > 0$. This is known as Hamilton's rule.

According to Wilson (1971), eusociality in insects has evolved once in the Isoptera and at least 11 times in the Hymenoptera. Why has social evolution occurred so frequently in the Hymenoptera? Hamilton (1964) presented a hypothesis based on the haplodiploid sex determination system in the Hymenoptera.

In haplodiploid species, males originate from unfertilized eggs and are haploid, while females originate from fertilized eggs and are diploid. A daughter therefore has half of her mother's genes and all of her father's genes. Thus, the coefficient of relatedness between a mother and a daughter is $1/2$, while that between full sisters is $3/4$, because all the daughters must share all the genes from their father. Thus, the daughters may 'prefer' to obtain the benefit of $r = 3/4$ by producing one more sister by an altruistic act, than to obtain the benefit of $r = 1/2$ by producing one more offspring.[2] This was considered to be a possible reason for the frequent evolution of eusociality in the Hymenoptera (Hamilton 1964; Wilson 1971). This theory is called the 3/4 relatedness hypothesis (West-Eberhard, 1978*b*).

Although Hamilton (1972) noted that haplodiploidy may not be the only factor in the evolution of eusociality in the Hymenoptera (see Chapter 4), this hypothesis has been featured in many textbooks (see Andersson 1984).

3.3 Parental manipulation and mutualistic aggregation hypotheses

In 1974, Alexander presented a theory of evolution of eusociality. Contrary to Hamilton (1964), whose interest was focused on daughter wasps, Alexander considered social evolution from the viewpoint of mothers (queens). If a mutation causes a new trait, which leads a mother (queen) to force her first batch of daughters to remain on their natal nest and to help their mother, and the mother who bears this trait can leave a larger number of grandchildren than other mothers, then the frequency of the trait in the progeny generation will increase, and, as the generations progress, the manipulation becomes more and more complete, the final result being eusociality. This is Alexander's *parental manipulation hypothesis*.

In this hypothesis, a single mutation can lead to social evolution, whilst in the kin-selection hypothesis, the recipient of the altruistic act must have the same allele common by descent with the donor. A female wasp can manipulate the sex of her offspring by opening her spermatheca to release sperm and fertilize eggs, and can suppress the ovarian growth of her offspring by decreasing the supply of food. The suppression of ovarian growth by pheromones plays a role in the latter stage of social evolution.

The parental manipulation hypothesis can explain the evolution of eusociality in the termites. In this case, pheromones emitted by queens and kings suppress the metamorphosis of larvae of both sexes. Michener and Brothers (1974) also presented a similar idea.

Although Trivers (1974) argued that a genetic change which leads to a revolt of daughters against parental manipulation so that they can product their own offspring must spread within the population, in temperate areas first-brood females which emerge in the early summer may be unable to leave many offspring until the autumn, even if they have risen in revolt against parental manipulation and established their own nests. It may be more advantageous

---

[2] As workers may increase the number of sisters (not offspring of sisters) by decreasing the number of their own offspring ($r = 0.5$), the cost should be $1/2C$ and the equation for the worker evolution is $B/C > 1/(2r)$ (Hamilton 1972). The basic line of comparison between diploid and haplodiploid species, however, does not change.

in this case for daughters to benefit by inclusive fitness effects, remaining at their natal nest. However, this explanation can be applied only to temperate species.

In 1972, Lin and Michener presented a critical review entitled *Evolution of sociality in insects*. This was, so far as I know, the first strong criticism of Hamilton's (1964) theory. They argued that the following three conditions must be realized for the validation of the kin-selection hypothesis: (1) social insects must have, at the beginning of their evolution, a matrifilial (mother–daughter) colony system; (2) within-generation caste system must be an abnormal phenomenon, or only seen in inbreeding populations; and (3) transfer of individuals between colonies must be suppressed. There are, however, many examples of within-generation cast systems in outbreeding populations, coexistence of multiple egg-layers within a colony, and transfer of females between colonies (even in the honey-bee, *Apis mellifera*, Sekiguchi and Sakagami (1966) described a translocation rate of 16.7 per cent of females). Worker reproduction is a common phenomenon in many eusocial bees and wasps.

Thus, Lin and Michener argued that the evolution of altruism cannot be explained by kin-selection alone, and that evolution through mutualistic aggregation must be emphasized much more. The subsocial wasps may first have aggregated to enjoy the benefits of aggregation (mainly for defence), and individuals belonging to the same generation may then have differentiated into within-generation castes, followed by the mother–daughter castes. I call this idea the *mutualistic aggregation hypothesis* (Fig. 3.1).

Although I do not regard haplodiploidy as an important factor in the social evolution of the Hymenoptera through 3/4 relatedness, it is true that

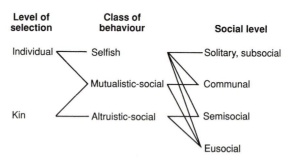

Fig. 3.1 Relations between modes of selection and social levels. According to Lin and Michener (1972), three classes of behaviour (selfish, mutualistic-social, and altruistic-social) have contributed together to social evolution, but the mutualistic behaviour which relates to individual selection might have dominated in areas or habitats with a high risk of predation.

haplodiploidy may serve as a favourable precondition for social evolution. Haploid males of the Hymenoptera may die when they have a lethal gene, and the population can therefore remove lethal genes effectively. This condition permits inbreeding, and also permits progeny to remain in their natal nest site (two characteristics commonly seen in the social Hymenoptera). A high level of relatedness among colony members as a result of haplodiploidy provides a precondition for the evolution of altruism (Hamilton 1972).

The three hypotheses explained in this chapter are, of course, not mutually exclusive (Andersson 1984).[3] The main aim of this book is to consider the relative roles of these hypotheses and to stress the importance of the third in the social evolution of the Vespidae.

---

[3] Seger's (1983) hypothesis on the evolution of eusociality in bivoltine adult-hibernating species has been omitted here because, although his theory is interesting, this is a special case of the kin-selection hypothesis.

4

Problems with the kin-selection hypothesis

4.1 Facts which contradict the 3/4 relatedness hypothesis

In this chapter, I will first discuss some facts which contradict the 3/4 relatedness hypothesis, and then introduce some studies of eusocial wasps made during the 1970s, the results of which have been considered to overcome one of the difficulties with the hypothesis. Although kin-selection may function regardless of whether or not 3/4 relatedness between sisters in the Hymenoptera has favoured social evolution, the application of general kin-selection theory is restricted by the factors discussed in this chapter. The social relations of some wasp groups are introduced here only as examples. More detailed accounts of the intranidal social relations of representative groups are given in later chapters.

The 3/4 relatedness hypothesis is based on the following three assumptions: (1) single insemination of a queen; (2) a mother–daughter caste system; and (3) haplometrosis (a single egg-layer (queen) per colony). Alexander's (1974) parental manipulation hypothesis is also based on the assumption of a haplometrotic, mother–daughter caste system. Thus, the two hypotheses postulate that the evolution of hymenopteran eusociality is a consequence of social evolution along Wheeler's (1923) subsocial route.

For assumption 1, there is only a little information about the number of matings in the eusocial Hymenoptera. Based on the ratio of the numbers of sperm a male has and the number that a queen receives during a mating flight, it is estimated that a queen honey-bee, *Apis mellifera*, mates with 7–17 males (Page 1986; for the significance of this, see Robinson and Page 1988, and Frumhoff and Baker 1988). In *Apis cerana*, Woyke (1975) estimated that a queen can mate with up to 30 males. Based on the fact that intranidal relatedness values are lower than expected in *Polistes metricus*, Metcalfe and Whitt (1977a,b) suggested that multiple matings take place in this species. I consider that multiple mating is not an abnormal phenomenon in eusocial bees and wasps.

In this respect, the 3/4 relatedness hypothesis in its original form might be abandoned as the major cause of the evolution of eusociality (Andersson 1984; see also Hamilton 1987; for criticism of the 3/4 related hypothesis using a population genetic model, see Kasuya 1982). Regardless of the validity of the 3/4 relatedness hypothesis, however, the evolution of eusociality is still easier when the worker and the recipients of the worker's help are sisters than when they are less related individuals.

For assumptions 2 and 3, Oster and Wilson (1978) considered that haplometrosis was the basic principle of the social organization of eusocial insects.

Wilson (1971) suggested that the single-queen colony is the basic social system of ants, because species with gigantic and well-organized colonies, such as *Eciton* and *Dorylus*, are haplometrotic. However, recent studies (Höll-dobler and Wilson 1990) suggest that pleometrosis is more common in ants than was previously suspected (it is thought to occur in more than 40 per cent of Japanese ants (K. Yamauchi, personal communication); see Chapter 11). Even in termites, which have previously been considered to be almost exclusively haplometrotic (single queen and single king), the number of known pleometrotic species is increasing (Chapter 11).

All of the six known species of honey-bee (*Apis* spp.) have single-queen colonies.[4] However, colonies of the species *Melipona nigra schencki*, sometimes have multiple queens (up to three, without overt conflict among them; S. F. Sakagami, personal communication). The other stingless bees are known to be single-queen species.

Most of the eusocial wasps for which social structures have been studied in detail (all living in temperate areas), are known to be haplometrotic or secondarily haplometrotic. Almost all of the 60 species of the subfamily Vespinae are known to be single-queen species (for exceptions see below). In the subfamily Polistinae, many temperate species of the genus *Polistes* are haplometrotic, including all those in mainland Japan (Chapter 5).

Multi-female-founding is however, not uncommon. For example, *Polistes dominulus* (= *gallicus*) in southern Europe (Pardi 1942, 1946, 1948), and *P. fuscatus* (e.g. West-Eberhard 1969) and *P. annularis* (Strassmann 1981*b*) in North America are known commonly to establish their colonies by foundress association. Yoshikawa (1962) and Iwata (1971) considered that most tropical *Polistes* might be pleometrotic. Rau (1933, 1943) reported that many colonies

---

[4] Although honey-bee workers lay eggs after they become queenless, all of these eggs develop into males. However, in *Apis mellifera capensis*, laying workers produce female offspring by thelytoky. Using a genetic model Moritz (1986) suggested that the high probability of queen loss due to strong winds in the Cape Province, South Africa may favour thelytoky of workers.

of *P. canadensis* and *P. versicolor* in Panama, and *P. carnifex*, *P. canadensis*, *P. instabilis*, and *P. exclamans* in Mexico are attended by multiple foundresses. In Cali, Columbia, where the climate is subtropical due to the high elevation, West-Eberhard (1969) found that 98 per cent (n = 46) of *P. erythrocephalus* (*P. canadensis* in the original publication) colonies are established by foundress association. Although *P. stigma* in Sumatra (S. Yamane, personal communication) and *P. subsericeus* and *P. niger* in Brazil (Hamilton 1972) are said to be haplometrotic, Hamilton himself considered the above examples to be exceptional cases (Hamilton 1972).

Even in the Vespidae, Siew and Sudderuddin (1982) and Matsuura (1983) found that some nests of *Vespa affinis indosinensis* are founded by multiple queens (Chapter 10). Large, perennial colonies of *Vespula germanica*, which were introduced into Australia, are pleometrotic (Chapter 10).

S. Yamane (1985) found that colonies of *Parapolybia varia* in Taiwan are founded by multiple females. However, *P. varia* colonies in Japan may be founded by single females (S. Yamane, personal communication) since its congener *P. indica* is haplometrotic in mainland Japan (Sekijima *et al.* 1980).

In the neotropical genus *Mischocyttarus*, five out of six species studied are known to be multi-female-founding species. The only exception is *M. flavitarsis* in Arizona, USA, the northernmost part of the distribution range of this genus (Litte 1979).

In the genus *Ropalidia*, all of which are tropical and subtropical species, all the species studied except for two (*R. taiwana*, Iwata 1969, and *R. formosae*, Wenzel 1987) are known to have multi-foundress colonies. Although Akre (1982) wrote that *R. variegata*, *R. cincta*, *R. fasciata*, and *R. artifex fuscata* were reported to be haplometrotic, this could depend on the observations that a pleometrotic colony may change to a haplometrotic one due to the development of a dominance-hierarchy. Except for *R. artifex fuscata*, for which detailed observations have not yet been made, all of the above species found their colonies with multiple foundresses at notably high rates (Chapters 6 and 8). In addition, there are some swarm-founding species (subgenus *Icarielia*) which have multiple queens (Chapter 9).

The three species so far studied of the genus *Belonogaster* in Africa, that is, *B. juncea* (Roubaud 1916; Keeping and Crew 1983), *B. griseus* (Marino Piccioli and Pardi 1970), and *B. petiolata* (Keeping and Crew 1987) also, as a rule, found their nests with multiple females.

Now we come to a group of neotropical wasps, *Polybia*, *Stelopolybia*, and some others. Many of the species belonging to these genera are characterized by a multi-queen colony system. In some cases, more than 100 morphologically distinct queens coexist in a single colony (Richards and Richards 1951; Jeanne 1991; see Chapter 9).

Thus, assumption 3 of the 3/4 relatedness hypothesis, that of haplometrosis, cannot hold without a special mechanism.

Using electrophoresis, Lester and Selander (1981) obtained results which suggested that the replacement of queens was common in *Polistes exclamance* and *P. apachus-bellicosus*.

4.2   The dominance-hierarchy/functional haplometrosis hypothesis

An explanation put forward by many authors (e.g. West-Eberhard 1969) to fill the gap between the 3/4 relatedness hypothesis and the existence of pleometrosis is that, as a result of the dominance-hierarchy, the ovaries of subordinate foundresses may degenerate, and/or eggs laid by the subordinates may be eaten by the dominant foundress.

In 1942, Pardi discovered a 'pecking order' among adult wasps in the Italian species, *Polistes gallicus* (now *P. dominulus*) (see also Pardi 1946, 1948). In this species, nests are often founded by an association of foundresses (82 per cent, maximum number of foundresses = 5), but there is frequent aggression among the cofoundresses. The dominance relationship is almost linear and the top-ranking foundress (called the α-foundress hereafter) does not perform much extranidal work, stays on the nest most of the time, and oviposits most eggs. Conversely, subordinates perform extranidal work most of time, like true workers. Pardi showed that the subordinates have degenerated ovaries. These subordinates are expelled from the nest after the emergence of workers.

Pardi also found that the α-foundresses eat the eggs laid by their subordinates. This fact was studied in detail by Gervet (1964) and called 'oophagie différentielle'. In differential oophagy, most of the eggs laid by subordinates are eaten by the α-foundress, but the reverse is rare (p. 45; see Table 6.2).

Dominance-hierarchies have been found in two Japanese paper wasps, *Polistes chinensis antennalis* (Morimoto 1961a,b) and *P. jadwigae* (Yoshikawa 1963). But these are basically single-female-founding species. Detailed studies of dominance-hierarchy among cofoundresses of basically multi-female-founding species were made by West-Eberhard (1969).

According to West-Eberhard, the life cycle of *Polistes fuscatus*, a species from Michigan, USA, is as follows: in spring a single foundress starts to build a nest, and over several days or weeks two to six foundresses join the nest. These joiners are usually females which have emerged from the same nest as the initial foundress the previous autumn (they may therefore be sisters). In some nests, however, associations of unrelated females are seen. Although associating foundresses (cofoundresses) perform co-operative nest enlargement and feed the larvae, there is a clear dominance-hierarchy among them. The dominant foundress (usually the female who has initiated nest founding) frequently attacks the other foundresses, with strong darting movements, biting, and even stinging. The subordinate foundresses gradually adopt a special posture, crouching on the nest. This posture may function to reduce attacks by the dominant foundress. As in *Polistes dominulus*, the dominant foundress remains on her nest most of time, while the subordinates perform most of the extranidal work. When the subordinate foundresses return to the nest with food, the dominant foundress solicits the food, often aggressively, and after she has taken it, she attacks the subordinate female, which then performs extranidal work again. Subordinate foundresses can oviposit during the early stage of the colony cycle, but most of their eggs are eaten by the dominant foundress. The ovaries of the subordinate foundresses degenerate gradually and the colony changes from pleometrosis to functional haplometrosis. Thus, subordinate foundresses may produce some workers, but they do not produce future queens. Only α-females can leave future queens.

In the neotropical wasp, *P. erythrocephalus*, subordinate foundresses never oviposit (West-Eberhard 1969). Such a strict hierarchy is also observed in some North American species (e.g. *P. metricus*; Gamboa *et al.* 1978).

During the 1960s, Pardi turned his attention to tropical wasps. According to Marino Piccioli and Pardi (1970), interactions between cofoundresses of *Belonogaster griseus* are sometimes mild, but there is a linear dominance relation. Oviposition is by high-ranking females and there is differential oophagy.

As stated above, *Parapolybia varia* found nests by associations of females, but there is always only one principal egg-layer per nest, even when the number of females exceeds 100 (Shima-Machado and S. Yamane, unpublished). After the emergence of many workers, the subordinate foundresses are expelled by the workers.

Thus, many authors have considered that colonies which were founded by pleometrosis may ultimately change to functionally haplometrotic ones, with a single α-female functioning as the queen. High relatedness among females can be proved by this process, and there is no contradiction with the kin-selection theory.

However, not all multi-female-founding species, are functionally haplometrotic. Studies of wasps in the wet tropics, have shown that many tropical and subtropical wasps have multiple egg-layers or even multiple, distinct queens in their colonies. Jeanne (1991) lists many tropical polistine species which have multiple egg-layers. I have found examples of such pleometrosis in Okinawa, Panama, and Australia. The principal object of this book is to describe the results of these observations and to discuss the factors which relate to pleometrosis in the tropics.

5

Comparison of dominance relations and proportion of multi-female nests in the Polistinae

5.1 Intranidal dominance relations

In this chapter I will discuss data relating to the frequency of intranidal dominance relations among female wasps, and to the proportion of multi-female-founding in the Polistinae. More than half the data about dominance-aggressive acts were collected by myself and will be discussed in detail in the following chapters, but the purpose of introducing these data in this chapter is to give readers a general picture of the social relations observed in wasps. I will also discuss here the methods used for describing social relations.

The basic study methods were individual marking and direct observations. In my own observations, wasps were collected from nests using a small net or a vial, then slightly anaesthetized with carbon dioxide and marked with a soft pen containing oil-soluble coloured ink. (The sunshine-tolerant colour pen, 'Opaque color' made by Magic® is one of the most convenient soft pens to use.) I marked up to four points on the dorsal part of the wasps' thorax using five colours. Then, while sitting or standing in front of a selected nest, I recorded verbally the behaviours of the wasps on the nest, using a cassette recorder, and later transcribed them.

Behaviours considered relevant to dominance-hierarchies (dominance-aggressive acts) were divided into the following two categories:

1. Weak dominance acts—an individual darted towards another, but neither biting nor grasping by the legs occurred (dart and stop short) or, even when there was slight contact, the other wasp (opponent) continued her preceding act or changed her position only slightly.

2. Strong dominance-aggressive acts—an individual jumped at, and/or bit, the body of the opponent, or pulled her with the mandibles. Sometimes the

attacking individual mounted and shook the opponent's body. The opponent adopted a submissive posture (see Fig. 7.1), escaped to the opposite side of the nest, or flew away. Sometimes both individuals fell from the nest while grappling together ('falling fights'; West-Eberhard 1969).

I have used this classification of dominance acts throughout my studies, although other entomologists do not necessarily use similar descriptive criteria. Entomologists who have made observations of especially aggressive species tend to record only the strong dominance-aggressive acts.

Some authors (e.g. Marino Piccioli and Pardi 1970; Darchen 1976a) described an act of food sharing as a dominance act, in which a female which has remained on the nest mounts another female, which has just returned to the nest, and makes mouth-to-mouth contact with the latter (see Fig. 6.2). These authors conclude that most of the liquid food is obtained by the dominant female through this act. Although this is true, this act is also often performed regardless of whether the mounted female is a forager or not. Many nest members commonly perform this act simultaneously when they have been excited. I call this act a 'kiss' and, except when it clearly indicated the forced removal of food (strong aggression), I did not record this as a dominance act (see pp. 40–1 for a discussion of the possibility of bias due to this).

The frequency of intranidal dominance-aggressive acts among 18 polistine species is shown in Table 5.1 (for social relations among multi-female nests of Japanese *Polistes*, see the next section). The values are expressed as the number of acts per female per hour (number of acts per hour divided by the maximum number of females seen on the nest during the observation). Observations of more than 30 minutes were used to calculate means and standard deviations.

I will show later that the frequency of dominance-aggressive acts varies markedly between nests and between different developmental periods on a nest. Of course, large variations in the frequencies may exist between different local populations of the same species (West-Eberhard 1986; Turillazzi and Turillazzi 1985; see pp. 45–50). We must, therefore, be careful when stating that a species is 'aggressive' or 'peaceful' from the results of observations on only a small number of nests. In addition, the values cited in Table 5.1 are means; in some species, as shown by the large standard deviations, substantial variation exists in the frequency of dominance-aggressive acts. Notwithstanding this, Table 5.1 may provide a first step for analysing the dominance-hierarchy among different 'species'.[5] (Note that the standard deviation of the frequency of dominance acts in *Polistes canadensis* (Panama population), a typically aggressive species, is small compared with the standard deviations observed in some other species.)

---

[5] I use the term 'species' in this book to designate local populations.

It is clear from Table 5.1 that the frequencies of dominance-aggressive acts in *P. canadensis* and in an Italian population of *P. dominulus* are high during both the founding (pre-emergence) stage and the post-emergence stage. In post-emergence stage nests, the frequency of strong dominance-aggressive acts is especially high. However, dominance-aggressive acts are almost non-existent in the pre-emergence stage nests of *P. versicolor, P. humilis synoecus, Ropalidia revolutionalis, Mischocyttarus angulatus,* and *M. basimacula.* The frequencies of dominance-aggressive acts for these species are significantly lower than that of *P. canadensis. Ropalidia fasciata* in Okinawa, however, is in an intermediate position, with large internidal variance. The frequency value of unity (1.0) or more for the number of strong dominance-aggressive acts per female per hour seems to be a value which approximately separates the aggressive and peaceful species.

All the species which showed significantly lower frequencies of dominance-aggressive acts than *Polistes canadensis* live in tropical or subtropical areas (the reason why the Panamanian species of *P. canadensis* is aggressive despite the humid tropical climate of its habitat, and some special features of its social system are discussed in footnote 12, p. 108). This is the reasoning on which one of my main arguments in this book is based, that foundress aggregations are favoured even when the relatedness among cofoundresses is low due to the high risk of colony failure and early death of foundresses. In the tropics, in particular the wet tropics, the predation pressure on wasp nests, especially by ants, is quite strong. We will consider this when comparing survival rates of single-foundress- and multi-foundress-founding nests.

5.2 Proportion of multi-female nests

Table 5.2 shows the percentage of multi-female nests,[6] the survival rates of single-female and multi-female nests, and the proportion of multi-female nests with multiple potential egg-layers in *Ropalidia fasciata* and seven mid-American and one Australian polistine wasps.

The percentage of multi-female nests is notably high in all the tropical species, and, although the sample size is small due to low numbers of single-female nests, none of the single-female nests of the four Panamanian species

---

[6] It is impossible to evaluate exactly the proportion of multi-female-founding in tropical wasps. In temperate regions, if we observe multiple females on a nest in spring, we can conclude that the nest was founded by a foundress association, while if we observe multiple females on a small nest in summer, the nest might have been reconstructed by a queen and her workers. In the wet tropics, however, nest foundation and reconstruction may take place throughout the year, so we cannot distinguish a nest which has just been founded by a group of inseminated females from that reconstructed by queen(s) and her (their) progeny. In the tropics, where males emerge throughout the year, most of the progeny females may be inseminated. There might therefore be no distinct boundary between nest foundation, reconstruction, and swarming in the tropical social wasps.

Table 5.1 Frequency of intranidal dominance-aggressive acts among females of polistine wasps. n is the number of observations of more than 30 min., and the value in parentheses is the mean number of different nests observed.

Species	Locality	Pre-emergence period				Post-emergence period				Author
		n	Total number of dominance-aggressive acts/♀/h	Coefficient of variation (%)	Number of strong dominance-aggressive acts/♀/h	n	Total number of dominance-aggressive acts/♀/h	Coefficient of variation (%)	Number of strong dominance-aggressive acts/♀/h	
Polistes										
dominulus	Pisa, Italy	1	1.95			1	4.43			Pardi (1942)
canadensis	Panama	6(5)	1.36 ± 0.55	40	1.22 ± 0.58	2(2)	2.61 ± 0.65	25	2.24 ± 0.13	Itô (1985*a*)
versicolor	Panama	7(3)	0.09 ± 0.07**	78	0**	4(2)	0.88 ± 0.81	92	0.51 ± 0.71	Itô (1985*a*)
annularis	Texas, USA	13(13)	1.06 ± 0.91	86	—	4(4)	0.28 ± 0.17	61	—	Strassmann (1981*b*)
instabilis	Texas, USA	(10)					3.6			Hughes and
	Mexico	(10)					1.7–1.9			Strassmann (1988)
bernardii richardsi	Darwin, Australia	5(1)	0**	0	0**					Itô (1986*c*)
humilis synoecus	Brisbane, Australia	6(1)	0.05 ± 0.01**	20	0**					Itô (1986*a*)

Ropalidia										
fasciata										
revolutionalis	Okinawa, Japan	30(24)	0.89±0.79	89	0.11±0.16**	44(29)	0.77±0.55**	71	0.20±0.22**	This study
	Brisbane, Australia	12(4)	0.04±0.09**	225	0**	5(3)	4.86±3.68	76	1.67±1.52	Itô (1987b)
sp. nr. *variegata*	Darwin, Australia	2(2)	0*	0	0*	15(4)	0.86±0.62	72	0.52±0.38	Itô and Yamane (1992)
g. gregaria	Darwin, Australia	4(2)	2.85±0.86	30	0.18±0.21	13(2)	1.56±1.02	66	0.94±0.75	Itô and Yamane (1992)
variegata										
jacobsoni	Sumatra	1	0.59		—	2(2)	1.63±0.36	22	—	Yamane (1986)
Mischocyttarus										
angulatus	Panama	3(2)	0*	0*	0*	3(1)	1.90±0.44	23	1.74±0.58	Itô (1985a)
basimacula	Panama	3(3)	0.56±0.22*	39	0*	2(1)	2.60±0.07	27	2.51±0.13	Itô (1985a)
drewseni	Brazil	1			0.1 ('dominations')	>0.35				Jeanne (1972)
mexicanus	Florida, USA	1			('bitings')†					Litte (1977)
labiatus	Columbia	6			>0.5 ('bitings')†					Litte (1981)
Parapolybia varia	Taiwan	3(1)	0.54±0.19§			1	0.21			Yamane (1985)

* The difference between these values and corresponding values for the typically aggressive Panamanian species, *P. canadensis*, is statistically significant at the 5% level (Mann-Whitney 2-tailed *U* test).
** As above, for the 1% level.
† Frequency of strong aggression might be higher than these values.
§ The frequency was especially high on the second day of nest founding (2.54/female/h) and then decreased somewhat. This value is the mean of data obtained on the third, fourth, and fifth days.

Table 5.2 The percentage of multi-female nests, the survival rates of multi-female and single-female nests, and the number of potential egg-layers on multi-female nests. The numbers in parentheses indicate sample size.

Species	Locality	Stage	Percentage of multi-female nests	Survival rate until emergence of first progeny		Percentage of multi-female nests with multiple egg-layers	Number of females with developed eggs in ovaries*	Number of females dissected
				Single-female nest (%)	Multi-female nest (%)			
Ropalidia fasciata	Okinawa, Japan	Pre-emergence	45(103)	31(26)[a]	64(28)[a]	50(4)	2/5, 1/3, 4/8, 1/3	
		Post-emergence				80(10)	3/8, 1/6, 7/22, 2/4, 3/19, 2/7, 2/8, 3/7, 6/16, 1/11	
Ropalidia revolutionalis	Queensland, Australia	Pre-emergence	94(16)	—	—	0(7)	1/6, 1/3, 1/7, 1/4, 1/4, 1/4, 1/4	
		Post-emergence				67(9)	5/11, 1/4, 4/7, 5/15, 2/6, 1/4, 1/5, 3/7, 7/28	
Mischocyttarus angulatus	Panama		82(11)	0(2)	100(4)	100(4)	3/5, 2/3, 3/3;	6/10
M. basimacula	Panama		61(18)	0(5)	75(4)	83(6)	3/5, 3/4, 3/5, 1/3, 2/3;	2/3
*M. mexicanus***	Florida, USA		67(171)	62(76)	78(41)	—		
M. flavitarsis†	Arizona, USA		2.2(134)	55(73)	—	—		
M. labiatus§	Columbia		42.9(58)	39(31)	63(27)	—		
Polistes versicolor	Panama		80(20)	0(4)	96(16)	57(7)	2/3, 1/5, 2/2, 1/2;	4/6, 1/8, 6/10
Polistes canadensis	Panama		98(90)	0(2)	50(12)	—		

* Italic fractions show values for the post-emergence period. As I could not often collect all the females, the percentage of nests with multiple potential egg-layers is somewhat underestimated.

** From Litte (1979). Survival rate until 56 days.

† From Litte (1979). Survival rate until 63 days.

§ From Litte (1981). Survival rate until 30 days.

[a] Significant difference at the 1% level.

survived until the emergence of the first progeny. It is also notable that the multiple potential egg-layers (females with developed oocytes in their ovaries) coexisted in all the species described here, even after the emergence of progeny, except during the pre-emergence period of *R. revolutionalis* (see Chapter 8). Although we need to obtain data on oophagy before drawing any firm conclusions, the coexistence of females with developed ovaries suggests that multiple egg-layers may often coexist on a nest, which can reduce the average relatedness among progeny females.

Data for temperate species comparable to Table 5.2 are shown in Table 5.3. As shown here, the percentage of multi-female-founding nests is less than 2 per cent in most Japanese *Polistes*. An exception is 11 per cent in *P. jadwigae* in Osaka (Yoshikawa 1957), but this value is based on only 18 colonies. As the sample sizes of the cases shown in Table 5.3 are large, it can be concluded that the colony founding system of Japanese *Polistes* (7 species) is basically single-female-founding, and that multi-female-founding may be an abnormal phenomenon. Table 5.3 also shows that the survival rate of single-female nests is notably high in Japanese *Polistes*.

Table 5.3 shows that the proportion of multi-female-founding nests in North American *Polistes* is notably high in the five species studied (the number of species in continental USA and Canada is 13; Richards 1978*a*). Although *P. annularis* lives in the subtropics, *P. fuscatus* and *P. variatus* found their nests with a high percentage of multi-female-founding, even in northern areas (Toronto, Canada, and Iowa and Michigan, USA, respectively).

Of 29 *P. dominulus* nests studied by Turillazzi *et al.* (1982), 19 were founded by foundress associations. If their data are based on random sampling, the amount of multi-female-founding is 65.5 per cent. At Helson near Odessa, Ukraine, 16 per cent of *P. dominulus* nests were thought to be established by multi-female-founding (Grechka and Kipyatkov 1984; they did not describe their sampling method).

We have no data on the method of nest foundation in species living in central Europe. The multi-female-founding of polistine wasps is known in Europe from Pardi's classic work (Pardi 1942, 1946); the incidence of multi-female-founding in northern and middle Europe is not considered to be high. Hamilton (1972) wrote that *P. gallicus* (*dominulus*) in central France may be considered to found their nests by single-female-founding.

What is the reason for the high proportion of multi-female-founding in North American *Polistes*? Many papers (Gibo 1978; Strassmann 1981*a*; Strassmann *et al.* 1988) mention that predation by birds is one of the most important factors in colony failure. A single-foundress colony is doomed if the foundress is killed, but a multi-foundress colony may survive. Although no quantitative data are given, Klahn's (1988) paper suggests that a large proportion of *P. fuscatus* nests are destroyed by predators, mostly birds. Multi-female colonies reconstruct their nests at a higher rate than single-

Table 5.3 The percentage of multi-female-founding nests and nest survival rate in Japanese, North American, and European *Polistes*.

Species	Locality	Percentage of multi-female-founding nests and (sample size)	Survival rate (%)		Author
			Single-female nests	Multi-female nests	
Polistes					
chinensis antennalis	Honshu, Japan	0(70)	43–84		Miyano (1980)
c. antennalis	Honshu, Japan	0.9(114)	49		Kasuya (1981 and personal communicaton)
c. antennalis	Kyushu, Japan	1.5(325)			Hoshikawa (1979)
riparius	Hokkaido, Japan	1.3(388)	73		Makino (1981 and personal communication)
jadwigae	Honshu, Japan	0.9(114)	53		Kasuya (1981)
jadwigae	Kyushu, Japan	0.5(369)			Hirose and Yamasaki (1984)
jadwigae snelleni	Honshu, Japan		53		Miyano (1980)
	Hokkaido, Japan	0.1(132)			Makino (personal communication)
snelleni	Honshu, Japan	0(80)			Itô, unpublished

P. fuscatus	Toronto, Canada	43.3(113)	6.7	30.6	Gibo (1978)
fuscatus	Michigan, USA	60.9(46)			West-Eberhard (1969)
exclamans	Texas, USA	13.8(159)			Strassmann (1981*a*)
variatus	Michigan, USA	54.5(220)			Metcalf (1980)
metricus	Kansas, USA	35.1(74)			Gamboa (1978)
metricus	Kansas, USA	7.0(100)			Starr (1976)
metricus	Kansas, USA	8.8(91)			Bohm (1972)
annularis	Texas, USA	91.5(637)	ca. 20	ca. 80	Strassmann (1989*a*)*
P. dominulus dominulus	Florence, Italy	65.5(29)			Turillazzi *et al.* (1982)**
	Odessa, Ukraine	25.0(64)	34	89	Grechka and Kipyatkov (1984)**
nympha	Florence, Italy	14.2(225)			Cervo and Turillazzi (1985)

* Calculated from Strassmann (1989). Nest survival rate until worker emergence.
** Proportions were calculated from the original data, based on the assumption that samples were taken at random.

female colonies (Klahn 1988). This is quite different from the situation in Japan where, despite detailed studies on the survivorship of nests (Matsuura 1977; Miyano 1980), nest failure due to birds was found to be negligible. Only 2 out of a total of 162 *P. chinensis antennalis* nests which were described by Miyano as 'disappeared' could be attributed to predation by birds. Predation of flying adults by birds was sometimes observed (Itô, personal observation), but the incidence of this is probably very low in Japan.

In North American *Polistes*, foundresses and workers often attack other nests of the same species in order to consume larvae and pupae (Gamboa 1978, 1980). The single-female nests are a better target for such cannibalism. In *P. fuscatus*, females which have lost their nests through predation (or by experimental removal) often take over other nests. This is an important source of nest loss (19.8 per cent) among single-foundress colonies, but rare (2.2 per cent) in multi-female colonies (Klahn 1988). This condition is also different from the situation in Japan. Although cannibalism by conspecifics is an important factor in nest failure (Kasuya *et al.* 1980), the level of usurpation is not as high (Itô, personal observations).

The principal factor leading to colony failure in Asian *Polistes* is predation by *Vespa* spp. (Matsuura and Sk. Yamane 1984; K. Tsuchida, personal communication). *V. tropica*, which feeds its larvae almost exclusively with immature stages of *Polistes*, is most important in this respect (e.g. for *P. jadwigae* it accounts for 40.7 ± 22.2 per cent of incidents of colony failure: calculated from Table 5.4 of Matsuura and Sk. Yamane 1984). The intensity of predation by *V. tropica* is not dependent on foundress group size, because, (1) in temperate areas *V. tropica* attacks polistine colonies mainly after the emergence of workers (0.4 per cent for pre-emergence period versus 46.2 per cent for post-emergence period in *P. jadwigae*; Matsuura and Sk. Yamane 1990), and (2) *Polistes* adults cannot repel the attackers.

Although comparative studies of the survival of *Polistes* colonies and of foundresses in North America and East Asia are needed in order to draw firm conclusions, it can be said that single-female-founding relates to conditions in which the risk of nest failure is low.

From the data shown in Tables 5.1, 5.2, and 5.3, I conclude that the following three processes are important in the evolution of the pleometrosis in wasps.

1. Among the tropical and subtropical polistine wasps, there exist species in which the mean frequency of dominance-aggressive acts among cofoundresses is very low (*P. versicolor, P. bernardi richardsi, P. humilis synoecus, Ropalidia fasciata, R. revolutionalis, R.* sp. nr. *variegata, Mischocyttarus angulatus,* and *M. basimacula*). In some species the frequency is low even after emergence of progeny females (*P. versicolor* and *R. fasciata*).

2. In many nests of these species, multiple potential egg-layers coexist even after the emergence of progeny.

3. Among tropical and subtropical species, the ratio of multi-female to single-female-founding seems to be far higher than among temperate species. Although a high percentage of some temperate species, such as *P. fuscatus*, found their nests by foundress associations, the number of cofoundresses is not large (up to 9, but usually fewer than 5). Conversely, the number of cofoundresses often exceeds 15 in many tropical species (see below).

4. The survival rate of single-female-founding nests of tropical species is very low, but that of five Japanese species, which are almost completely single-female-founding, is notably high. These facts indicate the importance of studies on the social biology of tropical eusocial wasps.

It must be noted, however, that, due to the large variation in social characters within and between populations (e.g. for dominance-hierarchy, see pp. 45–50), the comparative data shown in Tables 5.1, 5.2, and 5.3 are not entirely convincing. These provide only a starting point for future studies.

In the following three chapters, I will introduce in detail the results of my own studies on the tropical and subtropical independent-founding Polistinae, and in Chapters 9 and 10, I will discuss the results of recent studies on other eusocial wasps.

6

Ropalidia fasciata in Okinawa, Japan: a species with flexible social relations

6.1   Life history

Ropalidia fasciata is a wasp that is common in South-East Asia, the islands of Okinawa, Japan, being the northern limit of its distribution range (Plates 1–4). In Okinawa, this species lives in grasslands, where tussocks of a large perennial grass, *Miscanthus chinensis*, grow scattered over short grass and herbs, and in sugar cane fields. The nests are hung from the leaves of *M. chinensis* or sugar cane, except in rare cases, when they are hung from the shoots of trees. I studied the social behaviour and social structure of this species in Okinawa-Hontô (the main island of Okinawa) for a period of five years. A total of 374 nests, 646 foundresses, and 188 progeny females were marked.

Females that emerge and are inseminated in autumn usually overwinter in a group on their natal nest. Even if the nest has fallen down during winter, the females generally aggregate on a *Miscanthus* leaf near the original position of the nest. This behaviour is different from that of many temperate species of *Polistes*, in which a large number of females emerging from different nests may overwinter in a single hiding place (e.g. slits in buildings, holes in trees, etc.).

From March to April, females which have overwintered on a nest or at a common site may found a common nest (Plate 1). Usually, however, an overwintering group splits into several smaller groups, each of which may found a separate nest. Although the females that emerge from and overwinter on the same nest are not necessarily sisters (because there might be multiple egg-layers in their natal nests) the mean relatedness among cofoundresses has not been proved to be low. (I could not evaluate *r* using electrophoresis because there was only one polymorphic enzyme among 10 enzymes tested.)

In contrast with some other multi-female-founding species, such as *Polistes fuscatus* (West-Eberhard 1969) or *Belonogaster petiolata* (Keeping and Crew

Plate 1 Multi-female-founding in *Ropalidia fasciata* from Okinawa, Japan. The white marks seen on the gasters of two foundresses (arrowed) indicate that they emerged from the same nest the previous autumn, overwintered on their natal nest, and founded a nest together.

1987), in which a single female initiates a nest and other females gradually join it, *R. fasciata* cofoundresses usually aggregate from the initial stage of nest foundation (Itô *et al.* 1985).

In *R. fasciata* more than half of the nests were founded by foundress associations (55 per cent, n = 251, 5 years)[7] with the multi-female nests founded earlier than single-female nests (Itô 1985*c*; Sk. Yamane and Itô, unpublished). The number of cofoundresses are sometimes far greater than in temperate multi-female-founding species (e.g. *Polistes dominulus* and *P. fuscatus*), in which the number rarely exceeds five. The maximum number of cofoundresses in my survey station was 22 in 1983 and 17 in 1985 (Table 6.1), and 25 at another station (Itô and Iwahashi 1987). Single-female nests may be founded by females which have been isolated from the overwintering groups by strong winds or through dominance interactions on their natal nests during winter. Itô and Iwahashi (1987) found that the distribution pattern of the

---

[7] The number of cofoundresses per nest was either counted during the night, when all the extranidal workers should have returned to the nest, or was based on continual observations of a nest on which all the foundresses were individually marked.

Table 6.1 Percentage of nests founded by multiple foundresses, and related statistics, compared with nests founded by a single foundress of *Ropalidia fasciata* in Okinawa, Japan.

Year	Percentage of nests founded by multiple foundresses	Maximum number of cofoundresses	Mean number of cofoundresses	Survival rate (%) of		Number of nests reconstructed			Number of satellite nests built by	
				Single-female colony	Multi-female colony	Single-female colony	Multi-female colony	Unknown	Single-female colony	Multi-female colony
1983	53.4(58)	22	4.3(58)	46.2(26)**	86.7(28)	0	4	1	0	1
1984	52.9(119)	13	2.6(119)	36.4(55)**	77.8(63)	1	3	2	0	1
1985	63.8(47)	17	3.3(47)	33.3(15)**	88.0(25)	0	8	9	0	3
1986	51.9(27)	12	2.0(27)	30.8(13)*	76.9(13)	0	1	0	0	0
Total	55.0(251)	22	3.2(251)	38.5(109)**	81.9(129)	1†	16	12	0	5

N.B. The sample size for survival rate is not the same as sample sizes for percentage of multi-female nests and of number of cofoundresses, because we took some nests for dissection. The sample sizes (numbers of nests) are given in parentheses.
* $P < 0.05$ and ** $P < 0.01$; χ^2 test.
† $P < 0.001$; Fisher exact probability test.

number of cofoundresses per nest did not fit the truncated Poisson distribution, mainly because of the large number of nests founded by a single foundress (1-class frequency). They suggested that two types of progeny exist; (1) the usual progeny, and (2) the 'pioneer', which could discover new nesting sites; the latter might contribute to the higher than expected frequency of 1-class.

In temperate polistine wasps the first-brood progeny are usually females, which are destined to become workers. The males emerge in summer, just before or after the emergence of the second-brood reproductive females. In *R. fasciata*, however, males emerge throughout the year, and the first-brood females can mate with them (Itô and Sk. Yamane 1985).

Although most of the first-brood progeny may remain on their natal nest with or without ovipositing, some of them may found new nests in early summer and produce female progeny, while others may produce female progeny on their natal nest after the disappearance of their foundress (Itô and Sk. Yamane 1985). In autumn (sometimes until December) each colony produces females which are destined to overwinter and to found nests the following spring.

6.2 Mean fitness of cofoundresses in relation to foundress group size

In order to compare the costs and benefits of single-female- and multi-female-founding, we must evaluate the fitness, that is, the productivity of reproductive females, per female in the two types of nest foundation. However, this is difficult and we were unable to do it in our own studies. Instead, we used the number of cells produced per foundress until early June (when the first progeny emerge) and from the middle of July to early August, as an index of fitness.

The results of a four-year study show that the number of cells produced per nest increases with an increase in foundress group size. Nests founded by larger foundress groups also produced first-brood adults earlier (at the end of May) than small foundress groups (Itô 1985c).

The fact that the productivity of multi-female nests was greater than that of single-female nests does not necessarily mean that the productivity of the individual foundresses was higher in larger groups. Some authors (e.g. Michener 1964) stressed that pleometrosis is inefficient compared with haplometrosis, because productivity per foundress (not per nest) decreases with increasing size of the foundress group (for example, see Turillazzi *et al.* 1982). Much of these data (reviewed by Röseler 1985), however, are taken from surviving nests, with no description of colony survival rates. In *R. fasciata*, the number of cells produced per foundress was found to decrease with increasing size of the foundress group when we used data for surviving nests (Fig. 6.1, middle). In this population, however, many nests failed due to predation by ants, strong wings, death of the foundress, etc. The survival rate until emergence of at least one progeny in multi-female nests was always

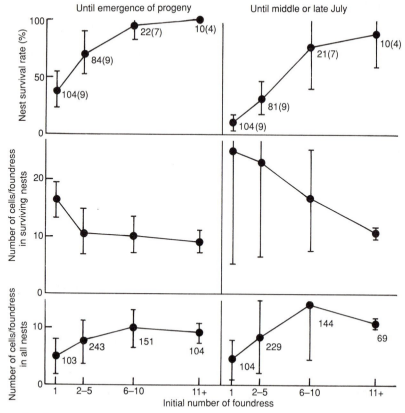

Fig. 6.1 Productivity (number of cells constructed per foundress) and survival rate of single-female and multi-female colonies until the emergence of the first progeny (end of May to beginning of June) and until middle or late July when most colonies reached the peak of progeny production. The numbers in the top figures show the number of nests studied and, in parentheses, the number of observed cases (product of numbers of survey stations and years). The numbers in the bottom figures show the number of foundresses at the initial stage of nest foundation.

significantly higher than that of single-female nests (Fig. 6.1, top), and there was a positive correlation between the survival rate and the foundress group size. We must therefore calculate the productivity of cells per foundress for all the nests which were established in spring (including the 'failed' nests).

As shown in Fig. 6.1 (bottom), the number of cells per foundress calculated for all the nests founded in spring first increased and then decreased; the maximum value occurring at a foundress group size of 6–10. Although the difference was not statistically significant, the same tendency was seen when data for each year were calculated separately. Provided that there is no difference between the 'effective number of eggs' (i.e. the number of eggs which survived oophagy) laid by each female on multi-female and single-female nests, multi-female-founding seems to be a better strategy than single-female-founding. Even when the subordinate females can lay half the number of eggs that dominant females can lay, and when there is no inclusive fitness effect, the fitness of females which have attended foundress groups of 6–10 females may be about the same as that which can be gained by single-female-founding. If there is an inclusive fitness effect, the subordinate females may have much larger fitness values.

In Okinawa, *R. fasciata* nests are often destroyed by frequent, strong typhoons or by attacks by ants. Although colonies can reconstruct their nests after such destruction, the frequency of successful nest reconstruction was found to be significantly higher in multi-female colonies than in single-female colonies (Table 6.1).

According to Gibo (1978), in *P. fuscatus*, multi-female colonies could reconstruct their nests at a much higher rate (0.43) than single-female colonies (0.07) (see Strassmann *et al.* 1988 for *P. bellicosus*). Litte (1981) showed that the nest survival rate of *Mischocyttarus labiatus* in Columbia was 39 per cent in single-foundress colonies and 63 per cent in multi-foundress colonies, and that the number of cells per foundress at the emergence of first progeny was larger in two-foundress and three-foundress colonies than in single-foundress colonies. The results of my observations on *R. fasciata* are basically similar to these results.

Although the females in an overwintering group of *R. fasciata* are not necessarily full sibs, the relatedness of cofoundresses, which are usually members of an overwintering group, may not be low. This provides some inclusive fitness gain for subordinates. In addition, they have the chance to become dominant during a colony cycle (see p. 108). Therefore, multi-female-founding in this species may be a good strategy for even subordinate foundresses, if they can effectively lay some eggs.

What are, then, the relations between cofoundresses?

6.3 Interactions among cofoundresses

I first gained the impression during field observations that social relations among cofoundresses of *R. fasciata* are far more 'peaceful' (mild) than reported among cofoundresses of other species (*Polistes dominulus*, *P. fuscatus*, etc.) (e.g. Itô 1983*b*). Although recent studies have shown that this is not always the case, I will begin this section with a description of the 'mild' relations and will then discuss variations in this behaviour.

In multi-female colonies of this species, two distinct types of foundresses almost always coexist; that is, queen-like foundresses, who rarely perform extranidal work and nearly always stay on the nest, and worker-like foundresses, who perform most of the extranidal work (the collection of honey, solid food, and pulp). This situation is the same as in all other eusocial polistine wasps so far studied. When I disturbed a nest, the foundresses who did not escape from the nest and who adopted alarm postures by expanding their wings, were usually the queen-like foundresses (they therefore perform a defensive role). The same behaviour was seen in a population of this species in Java (Turillazzi and Turillazzi 1985). In Okinawa, Japan, the number of queen-like foundresses per nest is one, two or, at most, three.

We observed the dominance-aggressive interactions among cofoundresses. Most of these interactions were performed by queen-like foundresses towards worker-like ones. On average, however, the frequency of dominance-aggressive acts was not as high as those of *P. dominulus* and *P. canadensis* (see Table 5.1). In addition, most of the dominance acts were weak, such as 'dart and stop short' or slight contact without any apparent change in the behaviour or position of the opponent. Biting, mounting with strong shakes of the opponent's body, and falling fights (West-Eberhard 1969) were usually rare.

We often observed the following behaviour. When a female returned to her nest from a foraging trip, a female (or females) who had stayed on the nest jumped on to the former's body. The latter, mounting the body of the returing female with her body well forward, bent over the thorax and head of the forager, attached her mouth to the returning female's mouth, and vibrated her body ('kiss'; Fig. 6.2; Itô 1983*b*; see p. 24).The kiss was, as a rule, performed when the donor had brought honey or water (the exchange of solid food was usually made with the food in contact with the mandibles of the two individuals). Similar behaviour was observed in *Ropalidia cincta* (Darchen 1976*a*) and *Belonogaster griseus* (Marino Piccioli and Pardi 1970). Darchen (1976*a*) and Marino Piccioli and Pardi (1970) described this behaviour as a dominance act. This is generally true in *R. fasciata*; that is, the queen-like foundresses are usually solicitors. The kiss is, however, sometimes made by subordinate foundresses towards the dominant foundress, and even by progeny females towards the foundresses which retain their dominant status (Itô 1983*a*). In addition, we often observed that several pairs of females performed this behaviour simultaneously, possibly stimulated by the kiss of other individuals. Therefore, I did not include these 'mild' kisses in the frequency of dominance-aggressive interactions (although I recognize that the kisses are often an expression of dominance status). But I recorded a kiss as a dominant act when the solicitor violently shook the opponent's body while the opponent crouched on the nest surface.

The problem might be that such omission of kisses from the repertoire of dominance acts could produce biases in the interpretation of social relations. However, as the aggressive species or populations (such as *Polistes canadensis* in Panama and *Ropalidia* sp. nr. *variegata* in Darwin, Australia) exhibit kisses

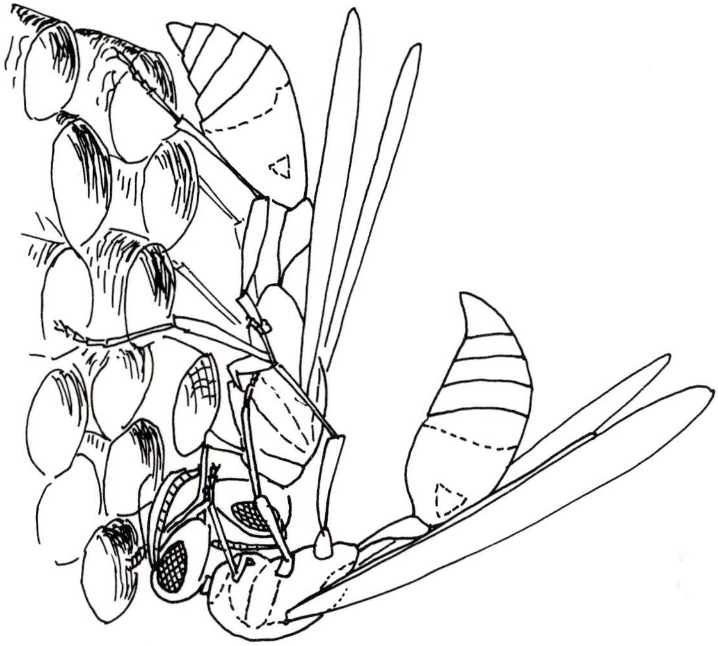

Fig. 6.2   The 'kiss' of *Ropalidia fasciata* females. This is usually initiated by a dominant female (right) jumping on to the body of another female who has just returned from a foraging trip (left). However, subordinate foundresses or even progeny females sometimes kiss the dominant foundress by jumping on to her body (Itô 1983*b*).

far more frequently than 'peaceful' species (e.g. *P. versicolor*), the variation in dominance relations among different species, as shown in Table 5.1, may not change when I add the number of kisses to the number of dominance acts.

   When too many foundresses aggregate on a small nest at the initial stage of nest foundation, I observed severe aggression between them. But the frequency of aggressive acts, as a rule, decreased with the growth of the nest, and a low frequency was maintained even after the emergence of progeny (Fig. 6.3).

6.4   Coexistence of multiple foundresses

It is known that in some polistine wasps the subordinate foundresses are driven away by the dominant foundress after the emergence of first-brood progeny (temporal pleometrosis; e.g. S. Yamane 1985, 1986 for *Parapolybia varia* and *R. variegata jacobsoni*). In *R. fasciata*, however, multiple foundresses often coexist even after the emergence of many progeny (Plate 2, Table 5.2 and

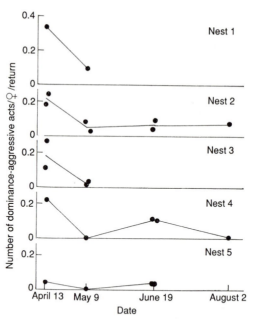

Fig. 6.3 Seasonal changes in the frequency of dominance-aggressive acts in five *R. fasciata* nests in Okinawa, Japan. At the initial stage of foundation of nests 1, 2, and 3 (April 13), many of the foundresses were unable to stay on their small nests due to overcrowding. There were many aggressive acts in these situations, but the frequency soon fell, and was maintained at a low level, even after the emergence of progeny (Itô 1985c).

Fig. 6.4). The mean survival rate of marked foundresses on their original multi-foundress nests until late July was 18.9 per cent (n = 166, see Fig. 6.4 for 1983 and 1984). Kendall's rank correlation coefficient (τ) between the survival rate and initial foundress group size was negative, but not quite significant, that is, -0.5 at the end of June ($0.05 > P > 0.01$) and -0.44 in late June to early August ($P > 0.05$). This suggests that, although there is conflict between the cofoundresses, the coexistence of multiple foundresses on a nest lasts until about two months after the emergence of the first progeny. Many of these surviving foundresses (76 per cent at the beginning of August 1983) had developed oocytes.

The absence of overt dominance acts and/or the coexistence of multiple foundresses, however, do not necessarily mean that the nest is truly pleometrotic, because it is possible that I did not recognize delicate, ritualized dominance interactions as reported by West-Eberhard (1982a) for *Polistes carnifex* (see later for a Javanese population of *R. fasciata*). However, if multiple females which have mature eggs in their ovaries coexist and

Plate 2 True pleometrosis in *Ropalidia fasciata* from Okinawa, Japan. The arrows indicate five foundresses that still coexisted after the emergence of many progeny females.

differential oophagy is absent or negligible in a nest, we can consider that the nest is truly pleometrotic, and the mean relatedness among progeny females should be low.

In Table 5.2, I showed that 50 per cent (pre-emergence) and 80 per cent (post-emergence) of the multi-female nests of *R. fasciata* possessed more than one female with developed eggs in the ovaries. For example, on a nest with 16 females (including progeny) collected on 3 August 1983, one female had five developed eggs (more than 0.8 mm in length), two had four eggs, three had three eggs, and ten had no developed eggs. The frequencies of oviposition and oophagy observed in *R. fasciata* and four other species are shown in Table 6.2. In *R. fasciata*, oviposition and oophagy were not strictly monopolized by the dominant foundress, in contrast with *P. dominulus*, *P. fuscatus*, *P. snellini*, and *P. annularis*. In *P. dominulus* and *P. fuscatus*, the α-foundresses lay more than 60 per cent of the eggs, and eat most of the eggs laid by their subordinates. In a

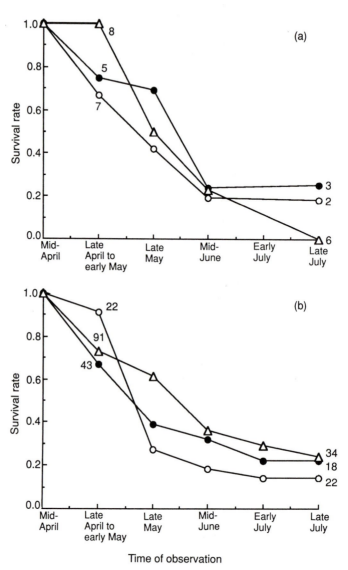

Fig. 6.4 The survival rate of marked foundresses of *Ropalidia fasciata* on their nests in relation to the initial foundress group size. (a) 1983, (b) 1984. The triangles show the results for initial group sizes of 2–5 foundresses, solid circles for 6–10, and open circles for 11–21. The results are given from Mid-April, just after nest foundation. The sample sizes (numbers on graphs) are different for different days, because the survival rate was calculated for foundresses whose nests survived.

Japanese species, *P. snelleni*, although workers can mate with males which emerge early and lay female eggs in orphan colonies (Suzuki 1985), oviposition in colonies with a queen present is done exclusively by queens (Suzuki 1987). Even in orphan colonies, top-ranked workers lay 85 per cent of eggs and eat most of the eggs laid by subordinates (Suzuki 1987). Although 'subordinate foundresses' lay eggs in *P. dominulus* and *P. fuscatus*, the subordinates which oviposit are β-foundresses only (Table 6.2).

6.5 Severe aggression observed in a satellite nest

The observations described above do not necessarily mean that the Okinawan population of *R. fasciata* cannot exhibit aggressive dominance-relations like those of *Polistes dominulus* or *P. fuscatus*.

In Okinawa, *R. fasciata* colonies sometimes make 'satellite nests' near the original nest during June and July (Plate 3; Itô 1986*a*). Such satellite nests are established with the co-operation of foundresses and progeny females, and, at least during the initial stage, there is frequent movement of wasps between the original and satellite nests. Thus, satellite nest construction is different from the process of swarming. A possible function of the satellite nests might be to spread the risk of destruction of the original nest due to typhoon or ant predation (see Strassmann 1981*a* for satellite nest construction in *Polistes exclamans*).

Colonies of Okinawan *R. fasciata* also in rare cases construct multiple comb nests (Plate 4; Kojima 1984*b*; Itô 1986*a*). The multiple comb nests are different from the satellite nests, in that the combs belonging to a single colony are all approximately the same age and all the established combs are used until the end of the colony cycle. The building of satellite nests and multiple combs indicates great flexibility of behaviour in this species.

Iwahashi (1989) recorded social behaviour among cofoundresses and progeny females on a satellite nest in 1984. Frequent severe aggression was observed among foundresses on the satellite nest, including falling fights. The dominance-hierarchy among females (including foundresses and first-brood progeny) was nearly linear. All the eggs laid by subordinates were eaten by the α-foundress, a situation quite different from my observations, shown in Table 6.2, and all the subordinate foundresses later disappeared. This situation is quite similar to the social relations observed for *P. fuscatus* and *P. erythrocephalus* (West-Erberhard 1969). Moreover, after the original nest was abandoned, the α-foundress was attacked by a progeny female in the satellite nest, and the nest was finally taken over by the latter.

Iwahashi (1989) also observed frequent aggression and differential oophagy by a dominant foundress on a nest during the initial stage of nest-founding. Observations were also made which were similar to examples found in an Indonesian population of this species (see p. 150). A nest was taken over

Table 6.2 Oviposition and oophagy by dominant (α) and subordinate females in some polistine wasp colonies. Numbers in parentheses are the mean number of eggs surviving until the end of the observation (>10 min).

Species	Stage	n	Oviposition by			Oophagy by			Author
			α-female	Subordinates	% by α-female	α-female	Subordinates	% by α-female	
R. fasciata	Pre-emergence	22	3	4(4)	43	0	1	0	Itô (unpublished)
	Post-emergence	22	11	10(8)	52	1	6	14	Itô (unpublished)
	Total	44	14	14(12)	50	1	7	13	Itô (unpublished)
	Pre-emergence	4	4	0	100	—	—	—	Kojima (1984*a*)
R. g. gregaria	Post-emergence	1	6	9	40	4	0	100	Itô and Yamane
	Post-emergence	1	3	2	60	3	2	60	(1992)
	Total	2	9	11	45	7	2	78	
P. dominulus	Pre-emergence	2	55	31†	59	31	7†	82	Pardi (1942)
	Post-emergence	2	131	69†	66**	164	36†	82	Pardi (1946)
	Total	4	186	100†	65**	195	43†	82**	
	Pre-emergence		41			41	6†	87**	Gervet (1964)
P. fuscatus	Pre-emergence	1	9	5†	64	4	2†	67*	West-Eberhard (1969)
P. annularis	Pre-emergence	13	25	17	60				Strassmann (1981*b*)
	Post-emergence	4	5	0	100*				Strassmann (1981*b*)
	Total	17	30	17	64				

* Difference between these and appropriate values for *R. fasciata* is significant at the 5% level; Fisher exact probability test (1-tail).
** As above, for the 1% level.
† By β-foundress only.

Plate 3 Original (O) and satellite (S, s) nests of *Ropalidia fasciata*. A small satellite nest (s) was soon abandoned.

(after severe fights) by the oldest progeny female, but she was later expelled by the second oldest female. Kojima (1984*a*) recorded the intranidal dominance acts among foundresses on pre-emergence stage nests. The frequency on a nest of eight females was initially low (0.7/♀/h on 8–13 April) but rose to a level comparable to that of *P. canadensis* (2.86/♀/h on 24 April), with the α-foundress initiating 69 per cent of the dominance acts.

I observed, however, that on another satellite nest (Plate 3), multiple egg-laying foundresses coexisted on 2 August, 40 days after the establishment of the nest, and that dominance acts were infrequent and weak. A foundress which did not leave the nest during my 2.5 h observation period, and which was also queen-like on 18–22 June, laid one egg, but it was eaten by another foundress which performed food collection on that day. Three eggs laid by two foundresses which also collected food were not eaten.

Thus, at least in Okinawa, intranidal social relations in *R. fascita* are characterized by great variability. Of course the social relations of eusocial wasps may be different on different nests and at different times; therefore, we cannot conclude too much from observations made on a small number of nests or for a short period. It is notable, however, that the standard deviation of the frequency of dominance-aggressive acts for *R. fasciata* was larger than that of *P. canadensis*, a typically aggressive species, and *P. versicolor*, a typically

Plate 4   A multiple comb nest of *Ropalidia fasciata*. A: large comb; B: smaller comb.

peaceful species (Table 5.1). The coefficients of variation of the frequency for pre-emergence and post-emergence periods was 89 per cent (range 58–128 per cent) and 71 per cent (range 33–89 per cent) respectively, for *R. fasciata*, compared with 40 per cent and 25 per cent for *P. canadensis*.

What factors are responsible for aggressive or peaceful interactions? Figure 6.5(a) shows the relation between foundress group size (for the post-emergence period, number of females on nest) of *R. fasciata* and the frequency of dominance-aggressive acts. There is a negative correlation in the post-emergence stage ($\tau = -0.36$, $P < 0.01$), but no clear relationship in the pre-emergence stage. This situation is somewhat different from the case of *Polistes annularis* reported by Strassmann (1981*b*), in which rank correlations between the foundress group size and the percentage of aggression performed by α-foundresses, and between the foundress group size and the percentage of eggs laid by α-foundresses are significant (-0.53 and -0.31, respectively; calculation by Itô). So far we have been unable to identify the major factor which contributes to the greater variation in the pre-emergence stage. There may be several reasons, such as relatedness, differences in ovarian conditions or in physical strength among the cofoundresses.

West-Eberhard (1982*b*, 1987) stressed that the behavioural flexibility of social insects is far greater than previously suspected, and that this great flexibility might be a factor which promoted their social evolution. She also

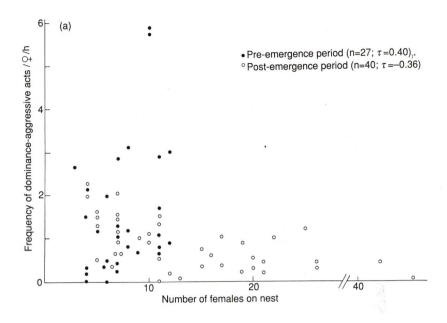

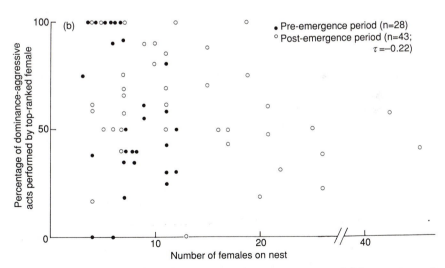

Fig. 6.5   (a) Frequency of dominance-aggressive acts on nests and (b) percentage of these acts performed by top-ranked *R. fasciata* females. Significant rank correlation was obtained between the frequency or percentage and number of females in the post-emergence nests (see text).

reported that the social behaviour of *P. canadensis* in Cali, Columbia, is different from that in Panama (see Chapter 7).

Turillazzi and Turillazzi (1985) reported on the social behaviour of *R. fasciata* in West Java. They observed that the first progeny female superseded the dominant foundress and was then substituted by a female who emerged later. They also reported a case in which the dominant foundress 'abdicated' her position to the newcomer without notable aggression (this suggests the existence of delicate dominance hierarchy interactions). In this population the kiss was always initiated by the dominant females. During 66 hours of observations, however, they did not observe any falling fights (Iwahashi observed falling fights, sometimes with attempts to sting in the Okinawan population; personal communication). They found that the higher ranked females had more developed ovaries (while they noted the coexistence of two females with 'very large ovaries' on a nest with 13 females). Two of the nine nests they observed were initiated by a single foundress but were joined later by other females. Thus, there may be marked differences in the behaviour of *R. fasciata* populations living in different places (in Java, the major habitat of *R. fasciata* is different from that in Okinawa; it hangs the nests from broad-leaved trees, rather than from grass).

When we use mean values, the social relations of the Okinawan population of *R. fasciata* in multi-female nests appear milder than those of *P. dominulus*, *P. erythrocephalus*, or *P. canadensis*, despite this species having a genetic trait to behave as a typically aggressive species (but see footnote 12 on p. 108 on the Panamanian species *P. canadensis*).

6.6 Summary

The Okinawan population of *R. fasciata* displays great flexibility in its social behaviour. Although it can behave as a typically aggressive wasp, the intranidal relations among foundresses are, on average, mild, and colonies are sometimes truly pleometrotic. Males are produced throughout the emergence season. Although first-brood females can found their own nests and leave female progeny, they usually remain on their natal nests. They usually behave as workers, but, at least sometimes, they can leave female-destined eggs on their natal nest, especially when the foundress disappears.

Multi-female nests survive far better than single-female nests, and, if the nests have been destroyed by a typhoon or by ants, they are reconstructed at a higher rate than single-female nests.

A foundress group size of 6–10 yields maximum productivity of cells per female, and this may relate to fitness. Notwithstanding this, severe aggression among females, monopolization of effective oviposition by the α-foundress, and repulsion of the foundress by the dominant progeny females are sometimes observed. The existence of notable variations in the intranidal social behaviour of this population is considered to be an adaptation to the special environmental conditions on Okinawa, with its frequent strong typhoons, high rate of predation by ants, and the ephemeral nature of its habitat—*Miscanthus* grasslands.

6.7 Appendix: some other *Ropalidia* species

The genus *Ropalidia* contains five subgenera including *Icariola*, *Anthreneida*, and *Icarielia*. In *Icarielia*, colonies reproduce by swarming and make enveloped nests. This subgenus is described in Chapter 9. Besides *Ropalidia fasciata*, the main subject of this chapter, and Australian *Ropalidia* (*Icariola*), which will be discussed in Chapter 8, species for which social behaviours or social structures were reported so far are as follows: *R. taiwana koshunensis* in Taiwan (Iwata 1969), *R. cincta* in Africa (Darchen 1976a), *R. marginata* (Gadgil and Mahabal 1974; Gadagkar and Gadgil 1978; Gadagkar 1980; Balavadi and Govindan 1981; Gadagkar and Joshi 1983) and *R. cyathiformis* (Gadagkar and Joshi 1982a,b, 1984) in India, and *R. variegata* in India (Davis 1966) and in Indonesia (S. Yamane 1986).

R. taiwana koshuensis builds nests of a very specialized type, a long comb with only one row of cells. Based on observations made of six nests, Iwata concluded that each nest was founded by a single foundress, and that this species is one of the most primitive *Ropalidia*; its social behaviour is somewhat similar to that of the Stenogastrinae.

According to Darchen (1976a), nests of *R. cincta* in the Ivory Coast are founded by single foundresses. In colonies which include many females, only the dominant females lay eggs. Non-ovipositing females often have degenerated ovaries, suggesting the impact of dominance-hierarchy on ovarian condition. These observations were consistent with those of Richards (1969).

R. marginata is a common paper wasp in India. Gadgil and Mahabal (1974) and Gadagkar and Gadgil (1978) reported that the nests of this species are founded by 1–20 females ($\bar{x} = 4.5$), and often last for more than one year. Each nest is first established by a single foundress which is later joined by additional females. This is the *P. fuscatus*-type of nest foundation. Nine out of ten small nests with less than 80 cells had a single potential egg-layer, while all of seven large nests with more than 100 cells had three to six females with developed ovaries. Mass emigrations of female wasps were observed when the nests became very large, suggesting that, although this is a basically independent-founding species, the colonies can sometimes reproduce by swarming (Gadagkar and Joshi (1985) described a process of colony fission in this species). Gadagkar (1980) and Gadagkar and Joshi (1983) studied intranidal social relations in two multi-female nests of this species. By using cluster analysis of frequencies of behaviours they recognized three categories of female wasps: sitters, fighters, and foragers. On each nest, only one of the sitters was the egg-layer (functional haplometrosis) and other sitters were considered to be hopeful queens or naïve workers. Fighters began to oviposit when the egg-laying sitter disappeared from the nest, or after their nests had reached a large size which allowed multiple egg-layers to coexist.

R. cyathiformis is another common paper wasp in India. According to Gadagkar and Joshi (1982a,b, 1984), both single-female-founding and multi-female-founding can be observed in this species. It often builds multiple combs, as do some Australian *Ropalidia* (p. 67), and three of ten females oviposited on a nest with three combs. A female which did not perform any extranidal work, however, laid the largest number of eggs. Cluster analysis again showed the existence of sitters, fighters, and foragers, but eggs were laid by fighters, not by sitters (Gadagkar and Joshi 1984; Gadagkar 1987). If the egg-layer was removed, one of the fighters began to lay eggs (Gadagkar 1987). Gadagkar suggested that the sitter of this species might be an idle or hopeful queen.

R. variegata is a common paper wasp in South-East Asia which builds slender nests (sometimes longer than 30 cm) which consist of two rows of cells. According to Yamane (1986), colonies of a Sumatran population (subspecies *jacobsoni*) are founded by single or multiple (2–4) females. The frequency of dominance acts in pre-emergence nests was, at least on a nest with two females, low ($0.59/♀/h$), but was high on two post-emergence nests (1.23 and $2.05/♀/h$). Dominance-hierarchy was still preserved even on a large nest with 20 females in which the α-female initiated 44 per cent of the dominance acts and in which at least five females were ovipositing. Conversely, only α-females oviposited on four nests in the early post-emergence period. Yamane found that in three nests the top-ranked foundress was ejected by either the first or second of her progeny to emerge. On one nest, the superseding female was again ejected by a female which emerged later (possibly a younger sister). As males emerge throughout the year, these progeny females may leave reproductive female progeny. This situation is similar to the 'serial polygyny' or 'short-term monogyny', reported by Jeanne (1972) for *Mischocyttarus drewseni*. Yamane (1986) concluded that the colony cycle of this species may proceed through a series of three stages; multi-female-founding, functional haplometrosis, and true pleometrosis.

Davis (1966) reported frequent intercolonial shifts of progeny females of *R. variegata variegata* in India. Yamane (1986), however, did not observe such a shift during his two-month observation.

Thus, the social structure of independent-founding *Ropalidia* (the subgenus *Icariola*) is quite variable, from functional haplometrosis to true pleometrosis. Most of them, however, found their nests by foundress association or swarms (groups of individuals belonging to more than one generation), and there is a notable tendency towards pleometrosis in this tropical genus.

7

Social relations in wasp colonies in the wet tropics: polistine wasps in Panama

7.1   A typically aggressive society: *Polistes canadensis*

In research in biochemistry or experimental biophysics, different researchers can obtain almost identical results by using similar apparatus for automatic chemical analysis or measurement of physical properties. Most descriptions of animal behaviour, however, are strongly influenced by the viewpoint and experience of the reasearcher (although there have been many attempts to standardize behavioural descriptions). Thus a social relation which I consider to be 'mild' may be described by another researcher as 'aggressive'.

In order to study the social relations of various tropical polistine wasps, I spent two months in 1982 conducting observations at the Smithsonian Tropical Research Institute on Barro Colorado Island, Panama (Plate 5). On and near Barro Colorado Island, I observed two species of *Polistes*—*P. canadensis* and *P. versicolor*—and two species of *Mischocyttarus*—*M. angulatus* and *M. basimacula*. I located nests of these four species and marked all the individuals on some nests. In other nests, only some of the individuals were marked. However, in both cases I could record precisely the frequency of dominance-aggressive acts among nest-mates.

The first species, *P. canadensis*, is a large *Polistes* with a dark-reddish-brown body and brown wings (Plate 6). This is the most common wasp species found on and near Barro Colorado Island. It suspends the nest under the eaves of buildings.

At the beginning of my observations, I was already aware that the social relations of this species are quite belligerent, with strong aggression among females on nests. This species is also most aggressive towards human beings, and often attacks the nests of other polistine wasps to take their larvae for food (see footnote 12 on p. 108). As shown in Table 5.1, the mean frequency of

Plate 5   Laboratories and bungalows on Barro Colorado Island, Panama.

intranidal dominance acts was 1.36/♀/h for the pre-emergence stage and 2.6/♀/h for the post-emergence stage. Most of the acts were of strong aggression (category 2, Chapter 5), that is one female jumped on another.

On being attacked by another female in this way, the opponent usually adopts a special posture, with her head down on to the nest and her abdomen raised (Fig. 7.1). This is considered to be a ritualized act which functions to decrease the attack (category 2).

Dominant females often violently vibrate their abdomen horizontally on nests, producing sounds which can be heard 2–3 m away from the nest. Abdominal vibrations have been reported in many other *Polistes* wasps and are thought to be a display to exhibit dominant status. For example, Gamboa and Dew (1981) observed *P. fuscatus*; in this species, horizontal vibration is

Plate 6 *Polistes canadensis* from Panama. Dominant females frequently dash at subordinates on their nests, and the latter often adopt special prostration postures (see Fig. 7.1). Dominant females often work on the nest, vibrating their gasters.

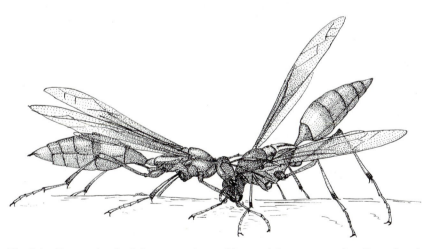

Fig. 7.1 Prostration in *Polistes canadensis*. The special posture a subordinate female assumes when attacked by a dominant female on their nest. This posture is considered to lead to a decrease in the intensity of the attack.

thought to be a form of communication between adults and larvae, while ventral vibration (beating the nest surface) displays dominance status. In *P. canadensis*, however, several females often perform horizontal vibration simultaneously on a single nest.

P. canadensis is closely related to *P. erythrocephalus*, which was studied by West-Eberhard (1969) in Cali, Colombia. The dominance-aggressive relations she observed in *P. erythrocephalus* nests were more severe than those of *P. fuscatus* (and included many falling fights). In contrast with the Panamanian population of *P. canadensis*, abdomen vibration in *P. erythrocephalus* is said to be performed by α-females only.

West-Eberhard (1987) studied a population of *P. canadensis* in Meta, Colombia; the results being similar to mine in some respects. She observed the absence of hierarchy among females on some nests. But detailed study showed that, on most established nests, despotic territorial females ('queens' in her terminology) suppressed aggressive behaviour in other members (this was realized not only by physical attack, but also by the 'approach' of the 'queen'). The frequency of dominance interactions was low in such nests. For example, in one nest with 17 foundresses the frequency was 0.17/♀/h, and nearly half of these interactions were initiated by the queen. Conversely, on another nest of about the same size and age, but lacking a despotic female, the aggressive interactions were 70 times more frequent (12/♀/h) and only 28 per cent of these involved the dominant female. Lateral vibration of the gaster was performed by several females.

West-Eberhard (1987) found that the 'queen' of Colombian *P. canadensis* became more aggressive when there were empty cells in the nest. Such queens physically chased other females from the lower 'shelf' region of the nests, which contained new cells, but when West-Eberhard tethered the queen to a different part of the nest, the subordinate females laid eggs. Neither the ritualized subordinate posture, as observed in the Panamanian population (Fig. 7.1), nor differential oophagy was observed. The removal of egg-layers induced oviposition by females of the next highest rank (thus many potential egg-layers coexisted on a nest, but their oviposition was suppressed by the dominant females).

An important finding in Panamanian *P. canadensis* is that unmated females are not aggressive, while mated ones are. As the queen can only suppress the behaviour of other females within close proximity, several females might be able to oviposit on the large, mature nests.

Pickering (1980), who first made a detailed study of the social biology of *P. canadensis* in Panama, found that some colonies established several nests up to 20 m apart. This 'extended colony' is different from the satellite nests of *Polistes exclamans* (Strassmann 1981a) and *R. fasciata* (see Chapter 6), and from the 'multiple comb' nests of a Brazilian population of *P. canadensis* (Jeanne 1979a). In the latter species, the distance between combs was several centimetres, thus this resembles the multiple comb nests often found in *Ropalidia revolutionalis* and *R.* sp. nr. *variegata* (see Chapter 8). Jeanne proposed that the multiple combs might be an adaptation to parasitism by a tineid moth, which was the major factor in nest failure in his study area. Pickering (1980) suggested that the trait of dispersing nests to form an extended colony system in Panamanian *P. canadensis* is an adaptation to reduce the risk of total loss due to attack by army ants, which is the most dominant predatory insect in the Panamanian tropical rainforests. For this purpose, the combs must be dispersed at distances of, at least, several metres. According to Pickering, the cofoundresses of a nest are usually sisters, but they often move to different nests, also established by their sisters.

Although the social interactions of *P. canadensis* in Barro Colorado Island were aggressive, the attacker was not restricted to α-females. If the group size was large, several dominant, aggressive females were seen. In Panama, the nests of this species become quite large (sometimes 30 cm long and 10 cm wide), and the nests frequently collapse. This may lead foundresses and progeny females to reconstruct their nests throughout the year. Thus, we cannot clearly distinguish nest foundation from nest reconstruction.

The social relations of Panamanian *P. canadensis* seem to be looser and more flexible than those of the Colombian population and of *P. erythrocephalus* in Colombia, in which only the α-foundress performs as a queen-like foundress. I believe that the existence of army ants, which are a quite influential predator in the Panamanian rainforests, is a selective force for such an especially flexible social structure.

I only observed single-female-founding in this species in Panama twice. These cases were of nests built in the small box of an electrical switch, and in a small hole in a tree; the size of these locations was inadequate for a large nest, and they were abandoned by the twentieth day.

7.2   A typically 'peaceful' social system: *Polistes versicolor*

Compared with *P. canadensis*, the intranidal social relations during the pre-emergence period of three other paper wasps I studied in Panama were 'peaceful'. In *P. versicolor* especially (Plate 7), which generally found their nests by association of females (80 per cent, n = 20), the peaceful intranidal relation extends into the post-emergence stage. Although up to three queen-like females were seen on every nest, and they performed alarm responses and attacked predators (including *P. canadensis*) which approached their nests, I did not observe any strong attacks on pre-emergence stage nests of this species (p. 26; Table 5.1). Dart and stop short behaviour, and sometimes a strong dash during the post-emergence stage, were observed. However, no attempts to bite or sting other females were observed. Many nests had multiple potential egg-layers (see Tables 5.2 and 7.1).

Plate 7 *Polistes versicolor* from Panama, showing individual marking. The intranidal social relations of this species are quite peaceful.

In this species I observed frequent shifting of females between two nests; again suggesting considerable flexibility of social systems (Table 7.1). Eight behaviour recordings (8 hours in total) and 17 point observations, made from August 30 to September 26 1982, where only the codes of individuals seen on each nest were recorded, showed that 9 of 11 females on the two nests shifted from one nest to the other at least once. Female number 021 changed her position 9 times. There were cases in which a female returned with solid food to a nest (nest 1) and gave part of it to the female(s) of that nest, and then flew to another nest (nest 2) and gave the remaining food to female(s) of the second nest. This fact indicates that shifting was not due simply to confusion. Although it is possible that the females of these two nests were sisters because of the short distance between them (50 cm), I observed another case of shifting in which a female moved to a nest which had been transferred from a point 200 m away; here the shifting female must have been unrelated to females of the transferred nest. Also, I observed that all the females of an abandoned nest (possibly the victim of an attack by army ants) moved to a nest 10 m away, and that the members of these two nests behaved normally.

All of these facts suggest that the intranidal social relation of *P. versicolor* is mild, and more flexible than some of the other species, such as *P. dominulus*, *P. fuscatus*, and *P. canadensis*.

Table 7.1   The shift of *Polistes versicolor* females between two nests (nest 2 and nest 3) and the number of observed ovipositions and developed oocytes in their ovaries (August 20–September 26 1982, Barro Colorado Island, Panama).

Females	Number (and number of days) of observations	Major place of sitting	Number of shifts between the two nests	Number of ovipositions observed	Number of developed oocytes (if female was dissected)
002	25(17)	Nest 2	2	2–3	10
003	25(17)	Nest 2	0		3
012	25(17)	Nest 1	2		—
031	25(17)	Nest 1	2		—
021	25(17)	Nest 1	9		0
041	25(17)	Both nests	2	1	—
006	25(17)	Both nests	3		0
066	16(13)	Nest 2	1		—
016	16(13)	Nest 2	0		—
042	14(12)	Both nests	5		5
Unmarked from nest 2	4(4)	Nest 2	1		4

7.3   Changes in the social systems of two *Mischocyttarus* species: 'peaceful' to aggressive society

The genus *Mischocyttarus* includes nearly 200 species. Most of the Panamanian species of this genus apparently live in the upper layer of tropical rainforests or suspend their nests from leaves or shoots in the undergrowth of the forest. Only *M. angulatus* and *M. basimacula* suspend their nests from the eaves of houses and other human constructions (in *M. angulatus* at least, I observed several cases in which the nests were suspended from leaves in the undergrowth). For this reason I concentrated my observations on these two species.

As shown in Table 5.2, at least 82 per cent of the nests of *M. angulatus* and 61 per cent of the nests of *M. basimacula* were multi-female-founded (since I was unable to count the number of females on some nests at night, these values are underestimates).

It was known that the nests of both species were founded by multiple females from the beginning of foundation (see Plates 8 and 9). Observations for more than 1 month from the initial stage of foundation did not show the addition of any new females to the nests. Although the sample size was small, due to the low rate of single-female-founding, all the single-female nests of both species failed (Table 5.2).

Plate 8 *Mischocyttarus angulatus* from Panama. This species regularly found their nests by association of females from the beginning of nest foundation. An ovipositing female (female 111 of Fig. 7.2; bottom) forages on the same day, and another female (top) stays on the nest. During the pre-emergence stage, there is no dominance behaviour among the foundresses.

I observed intranidal social relations in multi-female nests of *M. angulatus* for 13.5 hours (16 observations). I did not observe any dominance behaviour among the females. There were several cases in which females which stayed on the nest for a day went foraging on another day. On nest B8 (Fig. 7.2), female 006 did not forage on 17 September, but she performed two foraging trips while an unmarked female stayed on the nest most of the time on 29 September. On nest Q2 (Fig. 7.2), female 112 stayed on the nest on 30 September, while female 113 stayed on the nest on 4 October. In addition, female 111, which returned to the nest twice with honey and pulp on 30 September, laid an egg on 4 October (Plate 8). This egg was not eaten during the observation period. The numbers at the extreme right of Fig. 7.2 show that two females, 111 and 112, both performed extranidal work, and had developed oocytes, while 113, which did not perform extranidal work on 4 October, had no mature eggs. As nest Q2 had only 3 cells on 26 September, female 113 was not a daughter.

Such peaceful situations, however, changed dramatically after the emergence of progeny females. On nest B4 (Fig. 7.3), I did not observe any dominance-aggressive acts among three females 001, 002, and 003 on 30 August (female 001, however, took up an alarm posture at the approach of

Plate 9 Early stage of multi-female foundation in *Mischocyttarus basimacula* from Panama.

a *P. canadensis* female). Female 004 emerged from this nest on 4 or 5 September. On 7 September, female 001 went foraging and returned with solid food. Female 002 stayed on the nest most of the time, but foraged once for pulp. All four females performed extranidal work on 7 September (Fig. 7.3). Female 002 pecked 003 once, and the latter flew away. On 8 September, female 001, which stayed on the nest throughout my observations that day, pecked 002 once, then 002 flew away.

The aggression escalated on 17 September. I observed weak pecking acts four times during 1 hour, three of which were made by 001. Female 001 once violently attacked 004. On 22 September, however, female 001, which seemed to be dominant on 7 and 8 September, had disappeared, and 003 became a queen-like foundress. Only female 003 pecked other females (6 times).

I captured all the individuals on 22 September and dissection showed that three of the five females (two foundresses—002 and 003, and one oldest progeny, 004) had developed eggs in their ovaries.

After the emergence of many progeny, not only the frequency, but also the intensity, of aggression escalated (see Table 5.1). It was often observed that a female jumped on the body of another and bit her wings with her mandibles. Despite the attempts of the victim to escape, she pulled at the wings for more than 1 minute. In some cases, the victim became motionless, and hung from the mandibles of the attacker. The frequency of such strong aggression was not significantly different from that seen in *P. canadensis*.

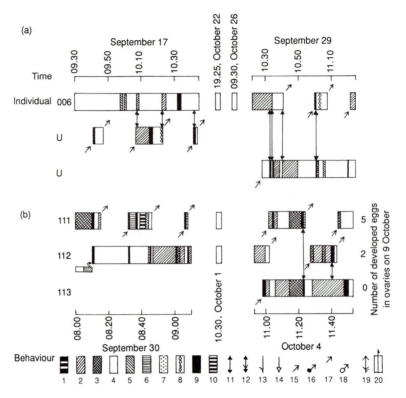

Fig. 7.2 Behaviour of *Mischocyttarus angulatus* females on (a) nest B8 and (b) nest Q2. 1, horizontal wagging of gaster; 2, resting on face of nest; 3, walking or grooming on face of nest; 4, resting on roof of nest; 5, walking or grooming on roof of nest; 6, cell construction; 7, mastication of flesh; 8, rubbing the gaster on pedicel or touching the pedicel with mouth; 9, insertion of head into cell; 10, oviposition; 11, kiss; 12, forced kiss (top individual was quite active while kissing); 13, weak dominance acts (top to bottom); 14, strong aggression (top to bottom); 15, return to nest without visible object; 16, return to nest with flesh; 17, fly away; 18, return to nest with pulp; 19, palpation by antennae (top individual was the initiator); 20, alarm posture in response to natural enemy. (After Itô 1984*b*).

In *M. basimacula* (Plate 9), I observed some weak dominance acts during the pre-emergence period, but the frequency was significantly lower than that of *P. canadensis* (Table 5.1), and I often could not determine which was the most dominant female. For example, both females 014 and U of nest F6 exhibited horizontal vibrations of the gaster on 13 September. On 19 September, female 012 twice exhibited weak pecking at other females and once a violent 'kiss', but female 012 carried on foraging and bringing back pulp. Dissection of females collected on this day showed that all three females had many developed eggs in their ovaries (Fig. 7.4). Three females which

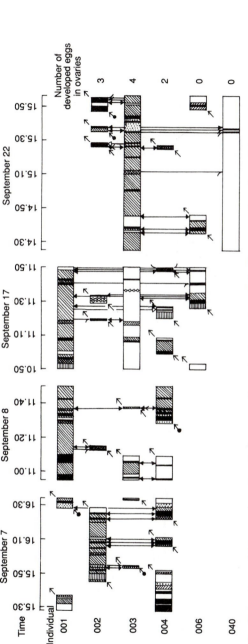

Fig. 7.3   Behaviour of *M. angulatus* females on a medium-sized nest (nest B4). For an explanation of the symbols, see Fig. 7.2.

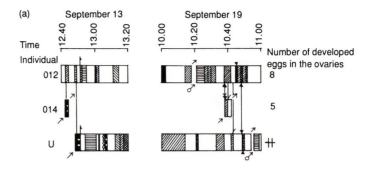

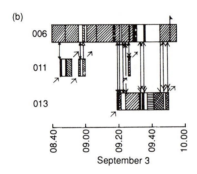

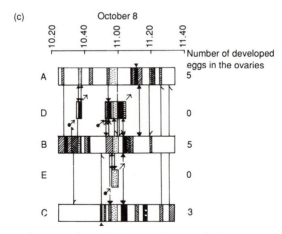

Fig. 7.4 Behaviour of *Mischocyttarus basimacula* females on small nests (a) nest F6, (b) B5, and (c) F4. + + indicates that there were several developed eggs but that they could not be counted. For an explanation of the other symbols, see Fig. 7.2 (Itô 1984*b*).

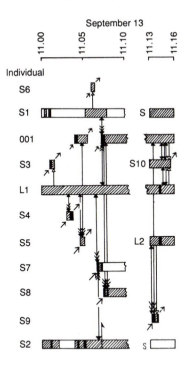

Fig. 7.5   Relationships among *M. basimacula* females on a large nest (nest F18). The position of each individual on the nest (striped area, face of nest; white area, roof side of nest) and occurrences of head insertion into a cell (black areas) are shown. Arrows indicate wasps returning or flying away as in Fig. 7.2.

stayed on nest F4 on 8 October had many developed eggs, while two females which performed extranidal work had none (see Chapter 1).

In this species dominance-aggressive interactions also escalated after the emergence of the progeny. An example is given in Fig. 7.5. It can be seen that five strong aggressions and 11 violent kisses (cases in which the opponent immediately flew away were scored as aggression) were observed during only a 13-minute period. This frequency was higher than that of *M. angulatus* (post-emergence stage) but there was no significant difference compared with *P. canadensis*.

Thus, it can be said for *M. angulatus* and *M. basimacula* that (1) intranidal relations among females during the pre-emergence stage were peaceful, while (2) severe attacks were frequently observed after the emergence of the progeny. I will discuss this escalation of aggression further in Chapter 12.

8

Role of multiple comb construction and the perennial nature of nests: polistine wasps in Australia

8.1 *Ropalidia revolutionalis*

In 1984, I had the opportunity to observe the social behaviour of polistine wasps in tropical and subtropical Australia. The Australian Polistinae consist of 12 species of the genus *Polistes* and 22 species of the genus *Ropalidia* (Richards 1978*b*), but there are only two species for which the social behaviour has been reported—*P. humilis* (Cumber 1951) and *R. revolutionalis* (Hook and Evans 1982).

R. revolutionalis is a common paper wasp from the east coast of Queensland, Australia. Hook and Evans (1982) observed this species in Brisbane, near the southern limit of its distribution range, and reported that its colony cycle begins with the foundation of a single comb but, as the colony cycle progresses, each colony builds additional combs near the original one (see Plate 10).[8]

Each comb consists of two rows of cells with a pedicel at one end. These features are similar to those of the nests of *R. variegata* (S. Yamane 1986) but, unlike *R. variegata*, in which each colony usually builds only one comb and its length often reaches 30 cm or more, the length of a comb in the nests of *R. revolutionalis* does not usually exceed 15 cm.

During the period September to October 1984 (the Australian spring), I studied 11 nests in Brisbane. At the beginning of October, nine of the nests had

---

[8] Construction of multiple combs seems to be a common feature of Australian *Ropalidia*; not only in *R. revolutionalis*, *R.* sp. nr. *variagata*, and *R. gregaria gregaria*, but also in at least two other unknown species. The term 'comb' usually means the separate layer of cells in a nest of, for example, hornets (Vespinae), which is covered by an envelope. However, I use this term here to designate separate nest units which hang side-by-side but are not covered by a common envelope (see Plate 10). This is the usage by Jeanne (1979*a*) and Hook and Evans (1982).

Plate 10 A multiple-comb nest of *Ropalidia revolutionalis* (photograph taken in R. L. Jeanne's study area in Townsville, Australia).

only one comb each, one nest had a developed comb plus a new comb with only one cell, and the last nest (1 comb) was an old comb which may have been built during the previous year. Except for the last case, the number of cells in each comb was 3–26 and there were no pupae. Thus, I concluded that September is the month of nest foundation for this species in the subtropical area of eastern Australia. All but two of the nests, which hung from two large, rotatable parasols on dining tables on a verandah, an abnormal habitat, were attended by multiple (2–7) foundresses.

All of the 17 nests I found at the beginning of September 1987 in Brisbane were also attended by multiple foundresses. Since even small nests with 2 or 3 cells had multiple females (some females which could not sit on the comb perched on twigs or eaves from which the nests were hanging), and I did not observe the recruitment of any new foundresses to the nests during the one-month observation period, I concluded that the type of multi-female-founding in this species is similar to that of *R. fasciata*, in which multiple females establish their nest co-operatively from the beginning of nest foundation.

Three out of 10 nests observed in Brisbane and all of five nests observed in Townsville in October 1986, had multiple combs, and all of these multiple-comb nests had evidence of adult emergence. Thus it can be concluded that the single-comb stage is approximately equivalent to the pre-emergence stage and the multiple-comb stage to the post-emergence stage of the colony cycle.

I observed intranidal social relations in single-comb stage nests for 16 hours (12 times on 4 nests), but I only recorded four weak dominance acts (see also Table 12.1). Thus, the social relations of this species during the pre-emergence stage were notably peaceful (p. 27, Table 5.1).

Despite this, dissection of foundresses taken from seven single-comb nests showed that each colony had only one female with developed eggs in the ovaries (Table 4.2; but, at least for three colonies, most of the females were inseminated). Thus, although colonies of this species are founded by multiple females, they soon become functionally haplometrotic, possibly through subtle dominance relations which I had overlooked. West-Eberhard (1982*b*) suggested such a subtle dominance relationship in *Polistes carnifex*.

The frequency of overt dominance acts, however, increased after the emergence of progeny (that is, at the multiple-comb stage) (Table 5.1; see also Table 12.1). In Townsville, where every stage of the colony cycle could be seen in October, the frequency on three developed, post-emergence stage nests was 2.95 ± 2.13 per female per hour. I often saw a dominant female expel a subordinate from the large comb, and sometimes drive her on to a small comb or its supporting structure (see next section). Such dominant females frequently moved from one comb to another. They often performed most of the food solicitation, adopted an alarm posture, and vibrated their wings on the comb. However, on developed, multiple-comb nests there were two or three dominant females.

Table 8.1 suggests that the colonies were functionally haplometrotic at the beginning of the post-emergence stage, but that they became pleometrotic when the number of females per colony increased. In fact, nests B612 and T601 had two females, each having more than ten developed eggs. Although I did not observe oviposition or oophagy during my 24-hour 45-minute period of observation, developed colonies possibly had multiple egg-layers (note that most of the females including progeny were inseminated; Table 8.1).

R. revolutionalis females overwinter (at least in part) on their natal nests, as does *R. fasciata*, and old combs were sometimes re-utilized in spring. Some of these 'perennial' colonies had more than ten combs.

8.2   Polistine wasps in Darwin: *Ropalidia* sp. nr. *variegata, R. gregaria gregaria,* and *Polistes bernardii richardsi*

Ropalidia sp. nr. *variegata* is a common paper wasp found near Darwin, in northern Australia. Like *R. revolutionalis*, this species constructs nests which consist of two rows of cells with a pedicel at one end. Each colony has a single comb at the pre-emergence stage but constructs satellite combs during the

Table 8.1 Number of potential egg-layers and inseminated females on nests of different developmental stages in *Ropalidia revolutionalis*.

Nest	Number of combs	Number of females dissected	Number of females with developed eggs	Number of females inseminated
1. Single-comb (pre-emergence) stage				
B402	1	6	1	—
B403	1	3	1	—
B404	1	7	1	—
B420	1	4	1	—
B609	1	4	1	3
B611	1	4	1	2
T605	1	4	1	3
Colonies with multiple potential egg-layers = 0%				
Inseminated females = 66.7%				
2. Multiple-comb (post-emergence) stage				
B601	2	11	5	10
B607	2	4	1	4
B612	3	7	4	6
T601	6	15*	5	10
T602	3	6	2	6
T603	3	4	1	2
T604	3	5	1	2
T607	3**	7	3	2
B699†	11	28	7	—
Colonies with multiple potential egg-layers = 62.5%				
Inseminated females = 71.2%				

   * Only half of the females seen on the nest were collected.
   ** Two combs were very small.
   † Near the final stage of the colony cycle, 77 males and 28 females were collected.

post-emergence stage (Plate 11). In August 1982, the mean number of combs per colony was 3.4 (n = 25, maximum 10). Nine of ten new nests I observed were founded by an association of females (but, as in Panamanian wasps, I could not determine whether a new single-comb nest was constructed by females of the same generation (foundresses) or was a reconstructed one built by females belonging to two generations).

As shown in Table 8.2, there were no dominance-aggressive acts on nests D5 and D8, which were at the single-comb stage. Conversely, the frequency was high (including strong aggression) on nest D9 and nest D2 on 27 October and 28. Most of the dominance-aggressive acts were performed by the α-females (90.5 per cent on 27 August and 92.3 per cent on 28 August in nest D2). Both were multiple-comb nests, and I found that emergence had just started on D2 at the beginning of my observations.

Plate 11 A multiple-comb nest of *Ropalidia* sp. nr. *variegata* from Darwin, Australia. Numbers show the possible order of construction of combs.

Table 8.2 Frequency of dominance-aggressive acts of four *Ropalidia* sp. nr. *variegata* and two *R. gregaria gregaria* nests in Darwin, Northern Territory, Australia.

Species						*R.* sp. nr. *variegata*				*R. g. gregaria*			
Nest			D2				D5	D8	D9	D11		D13	
Date	August	23	24	27	28	Mean	29	23	29	28	30	27	28
Number of females		7	8	10	8	8.25	4	3	7	8	8	5	5
Time of observed (h)		1	1.5	2	1	—	1.5	1	1.5	1.25	1	1	1
Number of kisses		2	6	11	8	—	4	0	10	2	1	0	3
Dominance acts*													
Weak		0	0	19	11	—	0	0	2	18	13	18	18
Strong		0	0	3	2	—	0	0	2	4	3	0	0
Total		0	0	22	13	—	0	0	4	22	16	18	18
Frequency/♀/h		0	0	1.10	1.63	0.68 ±0.72	0	0	0.48	2.20	2.00	3.60	3.60

* For categories of dominance acts, see p. 23.

The absence of dominance acts on 23 and 24 August suggests that the overt dominance act was almost lacking in the pre-emergence stage nests, as in *R. revolutionalis* (see Chapter 12). Nest D5, however, had two females with developed eggs.

In September 1987, I again observed social behaviour in this species on multiple-comb nests. The results were consistent with those observed in 1984, that is, the frequency of dominance-aggressive acts, including strong aggressions, on post-emergence stage nests was high (p. 27, Table 5.1).

R. variegata jacobsoni, an Indonesian species with adult morphology quite similar to *R.* sp. nr. *variegata* (J. Kojima, personal communication), does not construct multiple combs, but each colony constructs a single slender comb, sometimes more than 30 cm long. What is the reason for multiple-comb nest construction by *R.* sp. nr. *variegata* and *R. revolutionalis*? The distance between combs seems to be too small for it to be used as a counteradaptation to predation by ants (see p. 56). I did not observe any trace of moth parasitism, which was considered by Jeanne (1979*a*) to be a major reason for multiple comb construction in *Polistes canadensis* in Brazil.

One of the possible reasons for multiple-comb nest construction may be an adaptation to strong winds (which are quite frequent in Australia, compared with the absence of such winds in tropical rainforest areas). In this respect, Hook and Evans (1982) found a slender *R. revolutionalis* nest, 25 cm in length, within a forest where the nest was protected from wind. However, I propose another possibility: that is, the multiple combs might function to favour the coexistence of multiple egg-layers.

Table 8.3 gives examples of the position of each female on each comb of multiple-comb nests of *R.* sp. nr. *variegata* and *R. revolutionalis*. In *R.* sp. nr. *variegata*, the females which had the largest number of developed eggs (i.e. female 02 of nest D2 and female 15 of nest 05) (they are thus considered to be α-females), tended to occupy the largest comb, while those females without developed eggs were found on the small combs, shoots, wire, or wood from which the combs hung. *R. revolutionalis* exhibited a similar phenomenon (Itô 1987*b*). This fact raises the question as to whether the construction of multiple combs in this species (and in *R. revolutionalis*) permits the coexistence of multiple egg-layers within a colony. This might be a good subject for future study.

Ropalidia gregaria gregaria is a tropical species in which the size and colour of the adults and the shape of the nests are similar to those of *R. fasciata*. This may be basically a multi-female-founding species, because all three initial stage nests I found in August of 1984 were attended by more than five foundresses. As in *R.* sp. nr. *variegata*, this species constructs multiple combs (all three post-emergence stage nests found in September 1987 had multiple combs; Plate 12).

In Darwin, at least, intranidal social relations among the females of this

Table 8.3 Number of times each comb was selected by females.

Nest	Comb	Number of cells	Individuals								
R. revolutionalis											
B602			01**	02	03**	U's					Total
	Major	32	6	2	0	?					8
	Minor	30	5	4	9	?					18
	S*	—	0	0	0	?					0
	Number of developed oocytes		10	0	0	?					
B604			11**	05**	23**	U's					Total
	Major		4	1	1	4					10
	Minor		0	0	0	4					4
	S		0	3	3	0					6
	Number of developed oocytes		8	0	0	0,0,?					
R. sp. nr. *variegata*											
D2			02**	01**	03**	04	05	06	24**	U's	Total
	Major		68	58	21	25	15	16	27	120	350
	Minor		16	8	25	25	14	15	10	126	239
	S		4	0	0	3	2	10	20	11	50
	Number of developed oocytes		6	0	0	0	0	0	0	0,0	
D5			15**	21	22**	U					Total
	Major		20	6	0	5					31
	Minor		2	8	19	5					34
	S		0	1	3	1					5
	Number of developed oocytes		7	2	0	0					

* Found on shoots, eaves, or wire.
** Statistically significant difference from random distribution (5% level, Fisher exact probability test, neglecting values for S).

species seemed to be rather aggressive throughout the colony cycle. At the initial stage of nest foundation, I observed high rates of dominance-aggressive acts (2.85 ± 0.86/♀/h), a rate comparable to that of *P. canadensis* (Table 6.1). However, one of these nests had five potential egg-layers among nine females. The frequency of dominance-aggressive acts was the same in the post-emergence stage, although the intensity of strong aggression increased (0.91 ± 0.47 strong acts/♀/h).

Plate 12 A double comb nest of *Ropalidia gregaria gregaria* from Darwin, Australia.

Another species I observed in Darwin was *Polistes bernardii richardsi*, a species found only in tropical Australia. Table 8.4 shows the results of observations of a pre-emergence stage nest that had 32 cells and was attended by five females. Females 03 and 04 were seen on the nest most of the time, whereas females 01, 02, and 05 often left the nest and returned to it with food, liquid, or pulp. When a female returned to the nest with food or liquid, the

Table 8.4 Behaviour and ovarian condition of 5 *Polistes bernardii richardsi* females on a nest during 5 days (21–28 August 1984, Darwin, Northern Territory, Australia).

Females	Extranidal work	Return with			Oviposition	Vibration of gaster	Number of developed oocytes
		Liquid	Food	Pulp			
01	8			6		9	1
02	17		2	5			2
03	1					3	0
04	5			3	1	10	6
05	9	1	1	3			0

other females often solicited from it; I saw three transfers of food during the six hours of observation. During these six hours, I did not observe any overt dominance or aggressive acts on this nest, except for one case of a slight dart (from females 01 to 05) at 11:45, on 23 August. Horizontal vibration of the gaster, exhibited by dominant individuals of *P. erythrocephalus* (West-Eberhard 1969) and *P. canadensis* (p. 54), was performed by females 04 and 01, and, to a lesser extent, by female 03. Dissection showed that three of the five females had mature eggs in their ovaries. Because she was often seen near the pedicel and made frequent foraging trips, female 02 was considered to be a subordinate. However, her leaving the nest was not the result of any obvious aggression by the others. This female also had two developed eggs. Even female 04, a queen-like female, which had the largest number of developed oocytes and laid an egg during the period of observation, brought pulp to the nest three times, and was absent from the nest twice (for 15 minutes on 23 August).

Female 03 did not leave the nest during my observations, except for 3 minutes on 23 August. She cannibalized a larva on that day and an egg the following day. However, she had neither developed nor semi-developed eggs, and was unlike either a queen or a worker. She was possibly an 'idle' female, which suggests that we cannot assess a female as queen-like from the mere fact that she is always on the nest (see the discussion on *R. marginata* and *R. cyathiformis*, p. 51).

The social relationships of females in *P. bernardii richardsi* in the pre-emergence stage nest are much more peaceful than the typically aggressive species, such as *P. canadensis*; instead they resemble those of *P. versicolor*. However, I do not know whether multiple egg-layers continue to coexist after the emergence of the workers.

8.3 Three methods of nest foundation: *Polistes humilis*

Two subspecies have been discriminated for *Polistes humilis*: *P. h. humilis* is a common paper wasp in New South Wales, while *P. h. synoecus* occurs in Queensland. Cumber (1951) presented a pioneering paper on the sociobiological study of wasps (although it is not well known), based on *P. h. humilis*, which had been accidentally introduced and established in New Zealand. Cumber noted the reutilization of old nests and the coexistence of multiple potential egg-layers in this population. Douglas and Serventy (1951) studied a *P. h. synoecus* population, which was introduced into Western Australia. They remarked on the continuous utilization of nests for two years. In this respect, Richards (1978*b*) commented that 'although the reutilization of nests by social wasps is unusual, there was no reason to doubt these observations.'

In Brisbane, the colony cycle of *P. h. synoecus* begins in one of three different ways (Itô 1986*a*):

(1) foundation of new nests;

(2) reutilization of old nests by females which have overwintered on their natal nests (Plate 13); and

(3) continuous utilization of old nests (Plate 14).

Plate 13 Reutilization of an old nest of *Polistes humilis synoecus* from Brisbane, Australia. Overwintered females congregated on an old nest from which they had emerged the previous autumn. Most of them were laying eggs. The arrows indicate identification marks.

I distinguished category (2) from (3), when all the immature wasps of the previous year emerged or were abandoned by adults. In category (2), all the cells were empty during the winter and, in the spring, overwintered females began laying eggs in the old cells from which adults had emerged during the previous year. Compared with this, in category (3) large larvae, pupae, and adults overwinter together on the nest; we can therefore observe both mature and immature stages in early spring. In categories (2) and (3), many (sometimes more than 30) females initiate a new colony cycle. Conversely, most of the new nests were built by single females (Itô 1986*a* reported that all of six new nests were single-female nests, but I found a new nest in 1985 which was thought to be founded by several females).

On nest N1, which was a reutilized nest, I did not observe any dominance-aggressive act among more than 20 females (6.5 hours observation over 6 days), despite the fact that many females performed foraging activities. Even when I included such acts as non-violent clinging as weak dominance, the frequency of dominance-aggressive acts was only $0.05 \pm 0.01/♀/h$, which

Plate 14 Continuous use of an old nest of *P. humilis synoecus* from Brisbane, Australia.

was significantly lower than values for even *R. fasciata*. I could not detect any females which were notably dominant on the nest. However, four females made abdominal vibrations more frequently than the others.

Dissection of all females, collected after 22 days of observations, showed that more than 73 per cent (n = 19) of females had mature eggs in their ovaries. On another nest, N23, which had been used continuously, 94.7 per cent of females had mature eggs. All the females which oviposited (n = 6) did extranidal work, and oophagy was not observed. The reason why new nests were usually initiated by single females, despite such 'peaceful' relations among the females on old nests, is unclear.

8.4 *Ropalidia plebeiana*: a species which constructs 'towns' and divides the nest by biting it with the mandibles

Australia has a unique paper wasp species which forms huge aggregations of nests. Richards (1978*b*) described this as a new species, *Ropalidia plebeiana*, and reported on several nest aggregations he saw in 1972 in New South Wales. One of them, found under a concrete bridge over the Nelligen Creek, was a large aggregation at the time of discovery, while another, found under a bridge on Cabbage Tree Creek, was thought to be newly-formed. I observed these two aggregations in 1984, 1986, 1987, and 1989 (Plate 15); they had therefore persisted for at least 18 years.

The nest aggregation under Cabbage Tree Creek bridge (Plate 15) consisted of four subgroups and about 3300 nests (hereafter I refer to each nest as 'comb').

A question arises: whether such a huge aggregation of nests is a social unit like a 'supercolony' of some ants? A large supercolony of *Formica yessensis* found in a forest in Hokkaido (M. Ito 1973; Higashi 1976) consisted of several tens of thousands of queens and nearly 1×10^{12} workers.

We studied the behaviour of marked individuals on part of the nest aggregation under Cabbage Tree Creek bridge (Itô and Higashi 1987). The results showed that each comb, despite quite dense aggregation (Plates 15 and 16), was occupied by 1 to 13 females, and defended against approaching females by aggressive threat or fight behaviour by the owner(s). In early spring, eggs were laid in most of the old cells from which adults had emerged during the last year, and more than 90 per cent (n = 25) of females had been inseminated and about 70 per cent (n = 101) had mature eggs in their ovaries.

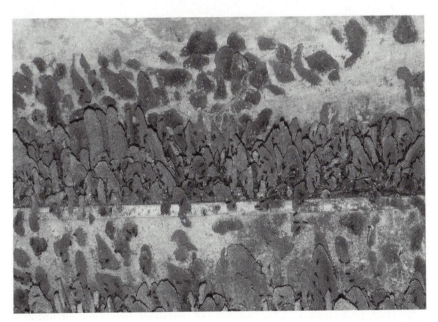

Plate 15 A 'town' of *Ropalidia plebeiana* under a bridge crossing Cabbage Tree Creek in the eastern mountainous area of New South Wales, Australia. This 'town' has persisted for at least 18 years since its discovery by the late O. W. Richards.

Thus, the nest aggregation of *R. plebeiana* is considered not to be a supercolony, but an aggregation of (at least initially) pleometrotic, independent nests.

This raises a new question. Why does *R. plebeiana* form such dense aggregations under bridges despite the fact that most of the undersurface of the bridge is vacant? The microclimate was not considered to be the main factor.

Plate 16 A close-up photograph of a *R. plebeiana* town in the final stage of the annual colony cycle in April 1988.

A possible reason is the 'selfish herd effect' proposed by Hamilton (1971), which may reduce the risk of central individuals being subject to predation. If there is a communal defence against natural enemies, however, each nest (as a social unit) is part of a loose social superstructure or nest aggregation. When I touched an individual nest in a dense colony, I was the subject of mass-attacks by a large number of wasps from surrounding nests. This is the first report of an insect society of multiple families with multiple social classes responding communally for mutual defence, and has previously only been reported for a few mammals, such as the gelada baboon (Kawai 1979).

Itô *et al.* (1988) counted the numbers of nests and attending females in a huge nest aggregation under a bridge near Bateman's Bay, New South Wales. The observations were as follows:

1. On 27 August 1987 there were 80 reutilized old nests.
2. On 13 November 1987 there were 188 nests (including 99 new nests), and 815 female foundresses.
3. On 10 April 1988 there were 191 nests, and 12 364 females, mostly new reproductives.

The survival rate of nests and the reproductive rate were notably high (about 90 per cent and a 15-fold increase, respectively), suggesting that the huge nest aggregations of this species might be beneficial for their existence, possibly through communal defence against predators and parasitoids.

R. plebeiana exhibits another striking behaviour. Before the appearance of the first larvae (in October), S. Yamane found that the foundresses of some large nests use their mandibles to cut the nest substrate, thereby dividing a single large nest into two or more independent parts. Thereafter, each separated part (nest) becomes a new social unit, because each nest was occupied by a particular group of females which, as a rule, were composed of a single egg-layer and several worker-like foundresses (Yamane *et al.* 1992). The increase in number of nests mentioned above includes new nests formed by nest division (50 on 13 November and 5 on 10 April). This is a completely new method of colony reproduction in social insects, and might be another unique way of permitting coexistence of many egg-layers. Thus, comb-cutting in this species may have a similar role to multiple comb construction in *R. revolutionalis*.

9

Multi-queen societies: swarm-founding wasps in the tropics

9.1 Neotropical swarm-founding wasps

Honey-bee colonies reproduce by 'swarming' (colony fission). When a colony produces new queens at a certain stage of colony growth, the old queen disperses from her nest together with a group of workers and this queen–worker group builds a new nest. Stingless bees also reproduce by swarming, but in this subfamily the new queens and some of the workers disperse from their natal nests (Sakagami 1984).

Many vespid wasps reproduce by swarming. In colonies of vespid species, in which swarm-founding is the rule, the nest consists of multiple combs covered by an envelope, made from pulp, or a mixture of pulp and leaves (Plates 17–20). Typical examples are seen in neotropical polistine wasps, the genus *Polybia*, and some other genera, especially those in which the generic names end in *-polybia*, such as *Stelopolybia* (now *Agelaia*, see Carpenter and Day 1988, and Carpenter 1989, but I will use the old name here), *Protopolybia*, *Metapolybia*[9], and the subgenus *Icarielia* of the Old World genus *Ropalidia*.

The swarm-founding wasps are notably different from the swarm-founding bees, at least in one respect, that is, each of their colonies tends to have multiple 'queens' (Jeanne 1980, 1991). (I use the term 'queen' here because at least some of the species discussed below have queens which are morphologically distinguishable from workers. In contrast, there is no distinct morphological difference between queens and workers in *Polistes*, *Mischocyttarus*, and the subgenus *Icariola* of the genus *Ropalidia* (all the *Ropalidia* species I referred to in the preceding chapters belong to this subgenus) discussed earlier.)

In their study of neotropical eusocial wasps, Richards and Richards (1951)

---

[9] According to Carpenter (1991), all of the New World polistine wasps (23 genera), except *Polistes* and *Mischocyttarus*, may be swarm-founding species.

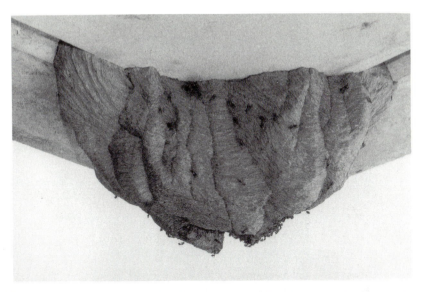

Plate 17   A nest of the swarm-founding paper wasp *Ropalidia romandi* hanging from the eaves of Griffith University, Brisbane, Australia.

Plate 18   The same nest as in Plate 17, showing the combs inside the envelope.

Plate 19 A nest of the Panamanian swarm-founding paper wasp, *Polybia rejecta*. This nest, which had been abandoned just before I collected it, probably had more than 10 000 workers and at least 100 queens.

described the relationship between morphology, ovarian condition, and insemination of females taken from the nests of a number of polistine species. In many species they found distinct morphological differences between inseminated females with developed ovaries and non-inseminated females with undeveloped ovaries. Considering that the former are queens and the latter workers, they found that, in many species, each colony contains 10 or more queens.

The morphological differences between queens and workers of '*Steropolybia' flavipennis* (Evans and West-Eberhard 1970) and '*S*'. *areata* (Jeanne and Fagen 1974) are illustrated in Fig. 9.1. In '*S*'. *areata*, there was no overlap between the characteristics of the two castes. Queens of '*S*' *flavipennis* have a colour pattern different from that of workers. Such clear differences were found in *Protopolybia sedula* (*pumila*) (Naumann 1970) and '*Stelopolybia*' *vicina* (based on photographs kindly shown to me by S. F. Sakagami).

Naumann (1970) observed the social behaviour of *Protopolybia sedula* in Panama. This species reproduces by swarming; the swarms consist of multiple queens and workers (and sometimes multiple males; Table 9.1). Examination of newly built nests showed that one nest was built by approximately 2000 workers and 113 queens, and another colony by about 1000 workers and 182 queens. This is an example of permanent pleometrosis, because the multiple

Plate 20   A nest of another Panamanian swarm-founding paper wasp, *Polybia scrobalis surinama*. There is an exit at the bottom-right of the nest.

queens coexist even after maturation of the colonies. Most of the queens had mature eggs in their ovaries. In seven of 14 nests which were collected for dissection, more than 80 per cent of the queens had mature eggs.

Plate 19 shows a nest of *Polybia rejecta* collected by myself in Barro Colorado Island. The nest consists of 17 combs under a hard envelope, 60 cm in length. Based on data given by Richards and Richards (1951) and Richards (1978*a*), we estimated that this nest had more than 10 000 workers and several hundred queens.

Plate 20 illustrates the nests of *Polybia (Myraptera) scrobalis* in Panama. In Costa Rica, Forsyth (1978) studied the social composition of a species of the same subgenus, *P. (M.) occidentalis*. It is clear from Table 9.1 that this species also has multi-queen colonies. Each colony produced many males and new queens when the colony had reached a certain age, and then swarming occurred. The size of swarming groups that Forsyth could detect was on average 237 (maximum > 540), including 2–21 ($\bar{x} = 7.8$) queens. When he observed the behaviour of marked wasps, by removing part of the envelope, 22 queens laid eggs during his 8.75-hour observation (Fig. 9.2). Machado (1985) studied a Brazilian species of the same subgenus, *P. (M.) paulista*. The number of cells per nest was 32 097–36 521 and the number of queens 18–478 (n = 9; Table 9.1).

Höfling and Machado (1985) reported the social composition of another

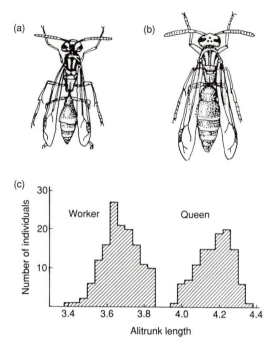

Fig. 9.1 Morphological differences between queens and workers of two paper wasp species belonging to the neotropical genus *Steropolybia* (now *Agelaia*). (a) A worker, and (b) a queen of *S. flavipennis*. (After Evans and West-Eberhard 1970, © The University of Michigan.) (c) bimodality in frequency distribution of wing length of *S. areata*. (After Jeanne and Fagen 1974.)

Brazilian species of *Polybia*, *P. ignobilis*. The numbers of combs and cells reached 5–13 and 6250–58 035, respectively. A large nest contained 90 queens which were considered to oviposit, 151 females having somewhat developed ovaries, and 3126 workers having undeveloped ovaries.

Carpenter and Ross (1984) reported on the social structure of some swarm-founding polistine wasps in Suriname. In *Polybia histriata*, the numbers of 'females' and 'queens' (i.e. inseminated females with mature oocytes—the authors did not describe any morphological differences) were 27 and 7 in nest A and 48 and 5 in nest B. In a nest of *P. catillifex*, they found 48 females, including 5 inseminated females with mature oocytes and 1 uninseminated female with mature oocytes. They found that a nest of *Brachygastra scutellaris* (this genus is known to collect and store large amounts of honey in their nests) which had at least 89 females contained 34 inseminated females with mature eggs and 22 uninseminated females with mature eggs.

The most surprising case of such multi-queen colonies is the composition of a nest of '*Stelopolybia*' *vicina*, collected by Sakagami and Zucchi in Brazil after

Table 9.1   Numbers of queens and workers in colonies of some tropical swarm-founding wasps. In *Agelaia areata, P. sedula, P. occidentalis, R. montana,* and *R. romandi,* queens (inseminated egg-layers) are morphologically distinguishable from workers. (Mean ± standard deviation, or when the sample size is 2, two values are shown.)

Species and stage	Number of queens	Numbers of workers	Number of males	n	Author
Agelaia areata	498 ± 190	5195 ± 1605		4	Jeanne and Fagen (1974)
Protopolybia sedula					
Development period	220 ± 172	7905 ± 5768	524 ± 1216	8	Naumann
New nests	71 ± 54	5593 ± 1191	1442 ± 989	3	(1970)
Swarming group	2 and 357	678 and 2822	—	2	
Polybia occidentalis					
New nests	21 and 20	296 and 224	0 and 0	2	Forsyth (1978)
Development period	57 ± 80	634 ± 750	0	3	
Polybia ignobilis	40 ± 35	1784 ± 841	0–934	7	Höfling and Machado (1985)
Polybia paulista	104 ± 150	1272 ± 836	0–337	9	Machado (1985)
Metapolybia azteca					
New nests	30 and 9	67 and 73	0 and 0	2	Forsyth (1978)
Development period	9 and 34	149 and 161	0 and 0	2	
Ropalidia montana	11 and 49	5189 and 8414		2	Yamane *et al.* (1983*a*)
Ropalidia romandi	13–29	ca. 4400		1	*Shima-Machad *et al.* (unpublished)

* J. Kojima (personal communication) gave values for two *R. romandi* nests; number of queens/total number of adults were ca. 50/1592 and ca. 141/2204, respectively (calculated by Itô). These data are cited in Jeanne (1991).

spraying the nest with insecticide (S. F. Sakagami, personal communication; Plates 21 and 22). It had 3000–4000 queens (which could be easily distinguished from workers by size and colour) and more than 1×10^6 workers.

Forsyth (1978) studied the social structure of *Metapolybia azteca*. This is a permanent multi-queen species, in which the swarms also contain multiple queens (Table 9.1). In contrast to *Metapolybia aztecoides* (see p. 90), many

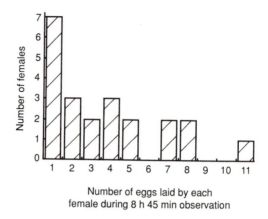

Number of eggs laid by each
female during 8 h 45 min observation

Fig. 9.2 Frequency distribution of individually marked queens on a nest of *Polybia occidentalis* in Costa Rica. During this observation 22 queens oviposited. (After Forsyth 1978.)

queens coexisted in colonies of this species until the later stages of the colony cycle.

At least some swarm-founding neotropical wasps may have a temporarily haplometrotic stage during their colony cycles, since single-queen colonies were reported in *Polybia ignobilis*, *Protopolybia scutellaris*, and some other, basically pleometrotic, species (West-Eberhard 1972; see section 9.4). But the basic social structure of many neotropical swarm-founding polistine wasps is considered to be the multi-queen colony system. Jeanne (1991) listed 38 neotropical polistine species; all of their colonies so far described are pleometrotic.

9.2 Swarm-founding *Ropalidia*: subgenus *Icarielia*

Plates 17 and 18 show a nest of *Ropalidia romandi*, collected in 1984 from a building at Griffith University, Brisbane, Australia. The length of the upper part was 40 cm and there were several dozen small combs under a thin paper envelope. During darkness, I covered this nest with a plastic bag, fumigated it with chloroform, and collected almost all of the colony members, which consisted of about 4400 females and 1400 males.

By using a cluster analysis of morphological characters, Shima-Machado and S. Yamane (unpublished) found significant differences between queens and workers (no worker had developed eggs in the ovaries) and estimated the number of queens to be 13–29. Multiple queens (> 100 in a colony) were also

Plate 21 A giant nest of the Brazilian paper wasp, *Stelopolybia vicina*. Another nest of similar size, collected by S. F. Sakagami and R. Zucchi by fumigating it with an insecticide, included 3000–4000 queens and more than 1 000 000 workers. The queens could be morphologically identified from the workers. (Photograph taken by S. F. Sakagami.)

found in two other nests which I collected in Brisbane and the Atherton Tableland in 1986.

Among the genus *Ropalidia*, species of the subgenus *Icariola*, which includes *fasciata*, *revolutionalis*, sp. nr. *variegata*, and *gregaria*, build non-enveloped nests and, as a rule, found their nests by individual-founding. Conversely, the subgenus *Icarielia*, which includes *romandi*, builds large, enveloped nests. The fact that in this subgenus queens are morphologically distinguishable from workers was first reported by S. Yamane *et al.* (1983a). In *R. montana* of India, the width of the first tergite of queens was larger (without overlap) than that of the workers (Fig. 9.3). The term 'workers' can be used here because all the smaller individuals had undeveloped eggs in their ovaries, while the larger individuals had developed eggs. The queen/worker ratio was 11/5189 in colony A and 49/8414 in colony B. S. Yamane *et al.* (1983a) supposed that the

Plate 22 A close-up photograph of part of the nest shown in Plate 19.

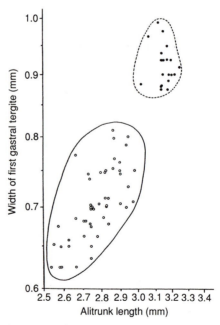

Fig. 9.3 Morphological diversification between queens (solid circles) and workers (open circles) in an Indian swarm-founding wasp, *Ropalidia montana*. (After S. Yamane *et al.* 1983.)

swarms of *R. montana* might contain multiple queens. A Sumatran species, *R. (Icarielia) flavopicta*, also exhibited queen/worker morphological differences and a multiple queen system. *R. (Icarielia) timida* may also found their nests by swarming (Pagden 1976). Thus, it can be assumed that all or most species of the subgenus *Icarielia* might reproduce by pleometrotic swarm-founding and permanent pleometrosis.

The genus *Ropalidia* consists of five subgenera, including *Icariola*, *Anthreneida*, and *Icarielia*. Aramaki (1985) found in *R. (A.) sumatorae* that one nest contained 1843 non-ovipositing (no mature eggs) females, 53 ovipositing females, and 8 males, and another nest contained 812 non-ovipositing and 605 ovipositing females. Most of the ovipositing females were found to be inseminated. This species produces multiple-comb nests in tree holes or other crevices (but without an envelope) and the colony size sometimes reaches several thousand individuals. Aramaki's data strongly suggest that this species is also permanently pleometrotic.

Even in the subgenus *Icariola*, some species make large perennial nests in holes and their colony size exceeds 1000. For example, Hook and Evans (1982) found a large nest of *R. socialistica* in Brisbane which had 1043 cells. I found several nests of this species in crevices of concrete buildings at Griffith University. I could not see the nests, but by counting the number of females seen on the wall of the building around the nest entrance (often more than 100), I estimate that these nests had more than 500 adults. At least some of them lasted for more than one year. Another possible case of true pleometrosis in the subgenus *Icariola* is *R. trichophthalma*; three colonies listed by Jeanne (1991) had 280–3288 adults, of which 7.5–22.3 per cent were 'queens', that is inseminated females with mature oocytes. A colony of *R. (Icariola) mackayensis* also had 118 queens among 590 females. I consider that some species, such as *R. socialistica*, are possibly permanently pleometrotic species in the subgenus *Icariola*.

According to Richards (1969) and Darchen (1976b), the African species, *Polybioides tabidus*, may found its nests by swarming and may possess multiple egg-layers (queens) in a single nest.

9.3 Relations among adult females on a nest

Does dominance order play a role in wasp colonies which have more than 100 egg-layers? We cannot assume the existence of a linear hierarchy among so many queens. Naumann (1970) wrote that queens of *Protopolybia sedula* formed a dense aggregation in the lower part of their nest. He did not describe any dominance-aggressive behaviour, despite detailed descriptions of other behaviours. I assume that he did not observe notably overt dominance acts.

Forsyth (1978) observed the nests of four *Metapolybia* species (*M. azteca*, *suffusa, docilis,* and *cingulata*) for 450 hours in total (he did not describe the individual observation times for each species), but he observed physical attack only once. He also wrote that the dominance behaviours of *Metapolybia* and *Polybia* were 'relatively mild and ritualized.' Simôes (1977, cited in Jeanne 1980) did not describe any aggressive behaviour in his detailed report on the social biology of '*Stelopolybia*' *pallipes.*

Queller *et al.* (1988) carried out electrophoretic analyses of colony members of the neotropical swarm-founding wasps, *Polybia occidentalis, Polybia serica,* and *Parischnogaster colobopterus*. Based on low mean relatedness values (0.34 for *P. occidentalis*, 0.28 for *P. sericata*, and 0.11 for *Parisch. colobopterus* (method of calculation given by Pamilo and Crozier 1982)), they estimated that each colony of *P. occidentalis*, *P. sericata*, and *Parisch. colobopterus* would have 2.62, 3.29, and 8.49 egg-laying queens, respectively. These values are smaller than expected, but if we consider the fact that the colonies of some species have far more than 100 queens, more analyses are necessary.

Although the evidence so far is scanty, these facts strongly suggest that the queens of at least some multi-queen species all have almost equivalent production of offspring; that is, the society is not held together by functional haplometrosis but by the communal coexistence of equipotent queens.

I mentioned earlier that a notable difference between swarm-founding species of bees and wasps is that the colonies swarm with a single queen or with multiple queens, respectively. However, there may be another difference. The queens of honey-bees and stingless bees cannot found their nests without the assistance of workers. This raises the question of whether the queens of swarm-founding wasps can found nests without the assistance of workers (as ant queens do). We have no data on this so far, but we cannot reject this possibility, at least in species in which the morphological differences between queen and worker are slight. The colonies of all or most of the wasp species mentioned in this chapter are perennial. This perennial nature and the resulting large colony size might be correlated with their multi-queen social systems.

9.4   *Metapolybia aztecoides*: co-operation or altruism?

Social relations among females of *Metapolybia aztecoides* were observed by West-Eberhard (1973, 1978a) in Colombia and Costa Rica, and seem to be somewhat different from the above-mentioned cases.

A female of this small paper wasp builds a flat, single-comb nest on a tree trunk or the wall of a house, and conceals it with a thin envelope. West-Eberhard made occasional observations of colonies at different developmental stages, and, in addition, detailed observations of social behaviours on a nest, by regularly removing part of the nest envelope. In total, 905 females were individually marked.

The colony cycle of this species (and of *M. docilis*; West-Eberhard 1973) begins with new nests being founded by swarms containing a number of 'workers' and one or many 'queens' (West-Eberhard's terminology). Although the queens could not be discriminated morphologically from the workers, they had developed ovaries (and therefore had extended gasters) and possessed sperm in their spermathecae. They gave a special display ('bending gaster') and often aggregated on one particular edge of their nests. Workers (non-egg-layers) did not perform the gaster-bending display; instead they performed a special dance when queens came to the centre of the nest, forcing them back to the edge again.

As shown in Fig. 9.4, an observed colony was initially pleometrotic (more than 30 egg-layers), but the number of queens decreased thereafter. This was also the case in some other colonies. The causes of this decrease were: (1) emigration of queens for swarming; (2) expulsion of some queens from the nest due to attacks by workers; and (3) change of status from queen to worker (remember that the status is determined by behaviour alone). Thus, before the emergence of progeny adults, colonies usually contained only one queen.

If the queen disappears, however, then many females begin to act as queens and commence ovipositing. In total 82 females were observed to lay eggs during West-Eberhard's study. These females (81 per cent of progeny females) emerged during a period roughly one month before the disappearance of the queen. This suggests that the presence or absence of queens may be a factor determining whether or not other females become egg-layers. After West-Eberhard (1973) removed the lone queen from a haplometrotic colony of *M. docilis*, she observed several females laying eggs and being treated as queens by the workers.

Thus, the colony shown in Fig. 9.4 changed from initial pleometrosis (February to August 1974) through haplometrosis (August 1974 to February 1975) to secondary pleometrosis (after March 1975). Whether secondary pleometrosis persisted until the end of the colony cycle or whether it again changed back to haplometrosis is not known. Although there is no direct evidence, all of the large colonies which contained adult males had multiple queens. I believe that colonies of this species may become permanently pleometrotic, at least in the later stages of the colony cycle (i.e. in large nests).

In such a colony cycle, a substantial proportion of the inseminated females, which were employed in the construction of the new nest, may later disappear from their nest, leaving behind only daughters destined to be workers. Is this altruism? West-Eberhard (1978a) considered that, as pleometrosis accelerates the development of nests, the subordinates might benefit from the inclusive fitness effect if they are sisters or cousins, and, in addition, they may have an opportunity to become egg-layers when the old queen has

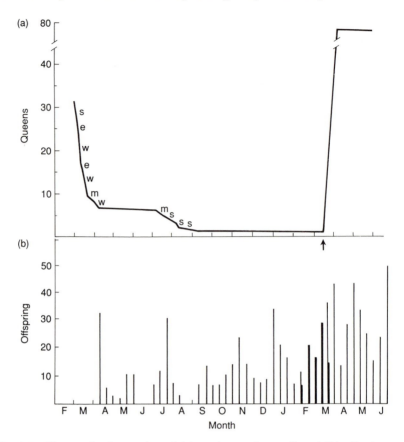

Fig. 9.4 Changes in the number of (a) egg-layers ('queens') and (b) offspring on a *Metapolybia aztecoides* nest during a period of 17 months after nest initiation. Although many queens coexisted at the initial stage of the colony cycle, many of them were expelled by workers (e), became subordinate, functional workers (w), or left the natal nest together with some workers to initiate their own nest(s), and thus the colony became functionally haplometrotic. If the sole queen died, however, young females (shown with thick vertical lines) began ovipositing simultaneously (after West-Eberhard 1978a).

disappeared. Thus, pleometrosis might develop through co-operative aggregation rather than through kin-selection. The fact that during the middle stage of the nest cycle the colony becomes haplometrotic may function to increase relatedness among new females.

Colonies of *Metapolybia azteca* are, according to Forsyth (1978), always pleometrotic, but those of *M. aztecoides* have a haplometrotic stage. I believe, however, that both *M. azteca* and *M. aztecoides* can perform both strategies;

functional haplometrosis and pleometrosis might be two extremes of their flexible social structure (recall the large variation in social behaviour in *R. fasciata*; Chapter 6). Such flexibility might be very important in elucidating the pathways of social evolution in wasps (West-Eberhard 1982*b*, 1987).

10

Social lives of the other social wasps

10.1   Pleometrosis and swarm-founding in the Vespinae

For a long time, the basic social system of the wasps and hornets of the subfamily Vespinae was thought to be exclusively haplometrotic (Jeanne 1980; Akre 1982). Although their nests all have multiple combs and envelopes similar to those of the nests of *Polybia* and some *Ropalidia* (subgenus *Icarielia*), no case of swarm-founding was known in this subfamily. Before 1983, besides rare, possibly exceptional, cases (see later), pleometrosis was found only in *Vespula germanica*, which had been accidentally introduced into Australia (Spradbery 1973). This case is an example of secondary pleometrosis, possibly a result of perennation of the colony cycle in Australia, where the winter climate is far milder than in Europe. Many entomologists, however, have considered this case to be an abnormal phenomenon resulting from the introduction of a small gene pool.

This once dominant viewpoint was questioned by the discovery of multi-female-founding and pleometrosis in a Sumatran hornet, *Vespa affinis indosinensis* (Plate 23), by Matsuura (1983). Eight of 11 colonies (73 per cent) discovered at the initial stage of nest foundation had multiple queens (in most Vespinae, in contrast to the Polistinae, queens can easily be distinguished from workers by their morphology). The number of queens in each nest ranged from 2 to 5 (mean = 3.1). Matsuura did not find any dominance-aggressive behaviour among nest-mate queens. In each pre-emergence stage nest, although all the queens collected had been inseminated, only one or two queens were considered to be egg-layers. In post-emergence stage nests, however, five of the six nests studied had multiple (2–5) queens, all of which had developed eggs in their ovaries (new queens were, if any, omitted from the analysis). Matsuura suggested that the social system of this population might be permanent pleometrosis; multiple egg-laying queens can coexist even after the emergence of several hundred workers. In contrast to these cases, colonies

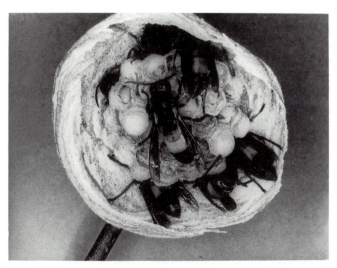

Plate 23 Early stage of a *Vespa affinis indosinensis* nest founded by multiple foundresses. (Photograph taken by M. Matsuura in Sumatra.)

of *V. affinis* in Japan and Taiwan are always founded by a single queen (Matsuura and S. Yamane, personal communication).

In fact, Rothnay had reported pleometrosis in *V. affinis affinis* in Bengal in 1877, but his report was ignored for more than a century (Spradbery 1991). Although Siew and Sudderudin (1982) found two cases of pleometrosis in Malayan *V. affinis*, and C. Starr found a case in *V. affinis nigriventris* in the Philippines (cited in Spradbery 1986), these reported only the coexistence of multiple queens.

Spradbery (1986) reported on the common occurrence of pleometrosis in a Papua New Guinean population of *V. affinis picea*. Of 19 colonies he observed from 1974 to 1984, 15 (79 per cent) had multiple queens. Spradbery divided the colony cycle of this species into the following three stages: (1) embryo colony (colonies with a queen, or queens, but no adult workers); (2) juvenile colony (colonies with adult workers, before the emergence of the new queen and production of males); and (3) mature colony (colonies with young queens and/or males present as immatures or adults). In the last stage, Spradbery could easily distinguish old queens from new ones by their worn, damaged bodies and wings. As shown in Table 10.1, multi-queen colonies outnumbered single-queen colonies at all colony stages, and the overall mean number of queens per colony was 3.5. Except for one colony with 15 queens, all the queens of these colonies were inseminated and had fully developed ovaries with at least one mature egg (or evidence of recent oviposition) in their ovaries. In the 15-queen colony, 87 per cent of the queens were inseminated and 11 had well-developed ovaries.

Table 10.1 Results of a population census of colonies of *Vespa affinis picea* collected near Port Moresby, Papua New Guinea. (From Spradbery 1986).

Stage of colony	Number of nests	Number of single-queen colonies	Number of multi-queen colonies	Mean number of queens/ colony	Maximum number of queens	Mean number of workers
Embryo	7	2	5	2.6	6	6
Juvenile	8	1	7	4.5	15	32.4
Mature	4	1	3	3.0	5	3402

Spradbery made a 10-hour observation of a pleometrotic embryo colony, which had six founding queens. During his observation, only one of the queens laid eggs twice; she made 15 foraging trips out of a total of 24 trips performed by all colony members (62.5 per cent). Except for one case of apparent food sharing, Spradbery found no overt interaction between the queens. These results were almost identical to those obtained by Matsuura (1983). Spradbery found a significant positive correlation ($r = 0.832$, $P < 0.001$) between the foundress group size and the number of cells per nest. In one pleometrotic embryo colony, the rate of cell-building was 0.3 cells per day, while in a pleometrotic colony, it was 0.7 cells per day. As all queens have the opportunity to leave reproductive progeny, pleometrosis in this species is beneficial for all the attending queens.

Secondary pleometrosis, in which new queens are recruited to their maternal nests, has been reported ten times in five species of *Vespula* (Spradbery 1986). *Vespula germanica* accounted for half of these reports (5/10), and all the cases involving *V. germanica* were discovered after the accidental introduction of this species to localities with a mild climate (e.g. Australia and South Africa). In Australia, *V. germanica* colonies become perennial and sometimes have more than 200 functional queens in the year following introduction, resulting in 1.2×10^6 cells and 120 000 workers (Spradbery 1991). No monopolization of oviposition by dominant individuals has been recognized in such huge colonies.

In addition, Matsuura found that swarm-founding is the method of colony foundation in the nocturnal hornet, *Provespa anomala*, in Sumatra (Matsuura and Yamane 1984). A group consisting of one queen and about 50 workers can establish a new nest. This is the first report of swarm-founding in the Vespinae. According to Matsuura, this may also be the case in two other species of *Provespa* (Matsuura and Yamane 1984). Matsuura found two embryo nests of *V. affinis indosinensis*, each of which was attended by queen(s) and an old worker. He considered that *V. affinis indosinensis* might also adopt swarm-founding (Matsuura and Yamane 1984), but, whether this is evidence of swarm-founding, or an example of reconstruction of a destroyed nest is not clear.

There is some doubt about whether pleometrosis as found in the Vespinae is a primitive stage in the social structure of this subfamily. In contrast to the Polistinae, the queens of all the vespine species can be morphologically distinguished from workers, and workers cannot produce female offspring. I propose that the worker-caste, which is clearly distinguishable from queens and cannot lay female eggs, could not evolve without kin-selection. If this is true, then pleometrosis in the Vespinae is a secondary trait which evolved from the original haplometrotic society. Even so, it is significant that typical pleometrosis in the Vespinae was discovered in tropical rainforest areas, where predation pressure is far higher than in temperate areas.

10.2 Social structure of the Stenogastrinae

The subfamily Stenogastrinae is a unique group among the Vespidae. Species of this subfamily do not fold their wings vertically like other vespid wasps (Carpenter 1982). They do not make solid peduncles (pillars) on their nests. Their nests are directly suspended from hanging fibres, such as the thin roots of plants. In contrast to the Vespinae and Polistinae, which build nests with finely crushed plant fibres, many stenogastrine wasps build nests with roughly crushed plant materials and soil, or almost exclusively with soil (Yoshikawa *et al.* 1969; Ohgushi *et al.* 1983; Hansell 1985*a,b*). This may be a trait which connects the Stenogastrinae with the Eumeninae, the 'mud-nest wasps'. In the Stenogastrinae, no morphological difference has so far been reported between foundresses and progeny females.

The social lives of the Stenogastrinae were first studied, more than 40 years ago, by F. X. Williams (1919, 1928).[10] Although Williams suggested that some species, at least, might be quasisocial or eusocial, no further study was made of this group until the end of World War II. A detailed paper by Yoshikawa *et al.* (1969) broke this long pause. They described the social relations of five Malayan stenogastrine species; three species of *Parischnogaster* and one each of *Liostenogaster* and *Eustenogaster*. Table 10.2 shows that in each of these species, multiple potential egg-layers coexisted when there were several females on a nest (but this does not necessarily imply multi-female-founding; see later).

For another species of *Parischnogaster*, *P.* sp., Yoshikawa *et al.* (1969) discovered a phenomenon which astonished hymenopterists around the world. There was a linear dominance order among the six females observed on a nest at Fraser Hill in central Malaysia. The top-ranked female never left the nest and was at rest or gently walked over the nest almost all the time (Fig. 10.1). Conversely, the lower the female's position in the dominance order, the more extranidal work she performed. Dissections showed that the higher

---

[10] Cited in Wheeler (1923) and Akre (1982).

Table 10.2 Frequency distributions of the number of females per nest and the number of potential egg-layers in four Malayan species of the Stenogastrinae. The numbers in the table indicate the number of nests. (Yoshikawa *et al.* 1969.)

Species	Number of females seen on a nest					Number of inseminated females with well-developed ovaries (multi-female nests only)*				
	1	2	3	4	5	0	1	2	3	4
Parischnogaster striatula	9	9	4	2	1	1+1**	2	8	4	0
P. alternata	1	2	4	2	0	0	2	2	3	1
Liostenogaster sp.	1	1	2	0	0	0	3	0	0	0
Eustenogaster ref. *fraterna*	4	1	0	1	0	0	1	0	1	0

* Number of females having ovaries at condition XX in Yoshikawa *et al.* (1969).
** On this nest, two females were inseminated and had ovaries of Xx.

ranked individuals had more developed ovaries and were inseminated. This relationship is quite similar to that found in *Polistes dominulus* by Pardi (1942).

Yoshikawa *et al.* (1969) found that females of *Parisch. striatula* often move from one nest to another nearby. Of 25 females observed, 20 moved one or more times over a period of 7 days (mean 1.4 ± 2.2, maximum 10 times). There was no relation between the frequency of internidal movement and ovarian condition. This fact, according to Yoshikawa *et al.* (1969), suggested that the larvae in a single nest were not necessarily sisters and brothers. Yoshikawa *et al.* (1969) suggested a 'loose' social order in stenogastrine societies, in which a nest may not be a social unit, although this opinion has been a subject of recent criticism.

According to Hansell (1981), nests of *Parisch. mellyi* in Thailand usually had one or two (rarely three) females and were actively defended against approaching alien females. In *Parisch. striatula* in Malaysia, the fate of alien females which approached or landed on the nests of other females was as follows: (1) they were expelled by the nest owners (5 times); (2) they landed once on an alien nest, but were later expelled by the nest owners which returned to the nests (3 times); (3) they landed on alien nests and ate eggs and/or larvae but were later expelled by the nest owners after their return (3 times); and (4) they were able to usurp the nests (5 times) (Hansell 1982). Thus, Hansell (1982) argued that the basic social unit of *Parischnogaster* is the nest-colony. In addition, Hansell raised doubts about West-Eberhard's polygyny-eusociality hypothesis (p. 2), because the social relations of at least some *Parischnogaster* species do not contradict Hamilton's rule.

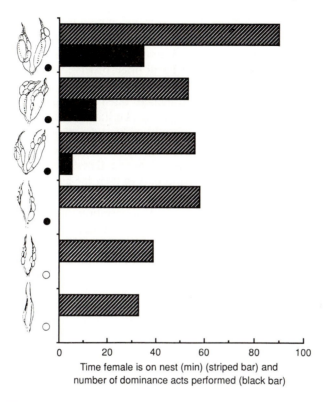

0 20 40 60 80 100

Time female is on nest (min) (striped bar) and
number of dominance acts performed (black bar)

Fig. 10.1 The relationship between dominance order and ovarian condition of
Parischnogaster sp. on a nest. Solid and open circles at the left of graph show
inseminated and uninseminated females, respectively (Yoshikawa *et al.* 1969).

According to Hansell's later work (1983), single-female-founding was the
rule in *Parisch. mellyi* in Thailand, but multiple egg-layers could coexist when
progeny females matured on their natal nests. In the latter case, older females
tended to remain on their nests most of the time and sit on the upper parts of
the nests. Dominance relations, however, were weak (0.11/♀/h; 34 h
observations).

Since 1980, Italian entomologists, including Pardi, have carried out
research on South-east Asian Stenogastrinae. According to their first report
(Turillazzi and Pardi 1982), the nests of Javanese *Parisch. nigricans serei* were
attended by 1–5 females. On one mature nest, all five females were found to be
inseminated. However, the number of females observed on young nests was
one or two, suggesting that single-female-founding might be the rule in this
species. They discovered many nests on which several males and a single
female were observed. At least some of these males came from other nests. The

attendance of so many males on a nest is uncommon in the Polistinae, but it has been seen in some other stenogastrine species. Linear hierarchy among females was observed on multi-female nests, and only high-ranked females had developed ovaries. Turillazzi and Pardi often observed strongly aggressive acts, such as biting and attempts at stinging.

It is known that second-ranked females often had yellow bodies in their ovaries—a sign of previous oviposition (Turillazzi 1987). Thus, the top-ranked female may suppress oviposition by subordinates. According to Turillazzi (1987), the first individual to emerge from a nest of *Parisch. nigricans serei* is usually a female which remains on her natal nest, helping her mother to rear the larvae. Many of them spend part of their life as foragers, indicating that this species is primitively eusocial.

Spradbery (1975) published a short description of the life history of *Stenogaster concinna* in Papua New Guinea. Since 31 of the 34 nests he examined were attended by only one female, this species probably founds nests with a single female.

S. Yamane *et al.* (1983) published a paper on the social behaviour of a Sumatran *Parisch. mellyi*. Although single-female-founding was the rule in this population, there was frequent internidal movement by females (Fig. 10.2) as in *Parisch. striatula* (Plate 24). In Fig. 10.2, female WB abandoned an old

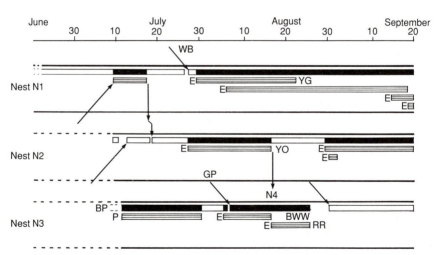

Fig. 10.2 Shifts of *Parischnogaster mellyi* females between nests. Arrows indicate an appearance on the nest or a shift to another nest (females which disappeared from the nest but could not found thereafter are not shown by arrows). Open, solid, and striped bars show that the female was solitary, dominant (did not forage) on multi-female nest, or subordinate (foraged), respectively. E, emergence. Note that newly emerged females worked for non-mother females, and a young female on N1 shifted to N2 during 18 and 19 July, and became an egg-layer on this nest.

Plate 24 A nest of *Parischnogaster mellyi*. A progeny female which has emerged from this nest sits on the top, while another immigrant female sits on the side of nest, (photograph taken by S. Yamane in Padan, Sumatra).

nest and founded a new nest G_2 (not shown), then moved to an orphan nest N1, leaving eggs and small larvae on G_2. Although female wasp YG, which could have been a daughter of the former nest owner, emerged on nest N1, she became a forager and reared the larvae laid by female WB, despite the apparent low relatedness between the two females. One week later, another female, a progeny of the former nest owner, emerged and also worked for female WB.

Nest usurpation was observed on nest N3. Although a dominant (BP) and a subordinate (P) were seen on this nest, BP disappeared on the day when a female (BWW) (possibly a daughter of BP) emerged. On the same day, a new female (GP) tried to land on nest N3, then BWW actively attacked GP, despite the fact that she had just emerged. After serious fights, GP was able to establish her dominant status on N3, and BWW (and RR, which emerged later) performed foraging activities for the immatures laid by the unrelated female GP.

These observations indicate the following: (1) in Sumatra, nests of *Parisch. mellyi* were defended by particular individuals (nest owners or females which had emerged from that nest) against intruders (as observed by Hansell (1983) in Thailand); but (2) there were frequent changes of nest ownership. One reason for these changes may be the high mortality rate of foraging females due to predation. Newly emerged females may at first work for their natal nests as subordinate workers, even when dominants are non-kin intruders, then found their own nests or move to orphan nests (see female YO in Fig. 10.2; she emerged on nest N2, worked on it for about 20 days, tried to found her own nest twice, but finally occupied an abandoned nest, N4 (not shown)). The fact that adult females belonging to two generations coexist and the progeny females, at least temporarily, work on their natal nests indicates that this species is at a primitive stage of eusociality.

From the results of dissections of females collected from the nests shown in Fig. 10.2, it was suggested that multiple inseminated females often coexisted on a nest (S. Yamane *et al.* 1983). Although single-female-founding is the rule in this species, inseminated progeny females can remain on their natal nests together with their mothers or even with usurpers. Oviposition may be virtually monopolized by the dominant female, but every female may have the ability to found her own nest and lay eggs destined to be females.

Thus, at least in *Parisch. mellyi* and *nigricans serrei*, nests are founded, as a rule, by a single female. Queen supersedure was often observed in *Parisch. mellyi* and *nigricans serrei* (Turillazzi 1987), and *Liostenogaster flavolineata* (Hansell *et al.* 1982). There are however, species which found their nests by foundress associations.

For example, in *Parisch. alternata* (see also Table 9.2), 41 of 46 colonies (89.1 per cent) studied by Turillazzi (1986) in 1984 were attended by two or more females (maximum 9, mean 3.17, calculated by Itô). Although some of these nests might be attended by a mother and her daughter, Turillazzi (1986) considered that some nests, at least, were established by foundress associations. According to Turillazzi, 84.7 per cent of females on multi-female nests were inseminated and 51.4 per cent had developed oocytes in their ovaries. More than half of the multi-female nests (57.8 per cent) had two or more potential egg-layers. For example, a nest attended by nine females had six potential egg-layers and another nest attended by the same number of females had four potential egg-layers. Potential egg-layers belonging to two successive generations (e.g. mother and daughter) can, of course, coexist on a nest.

Results of observations by Turillazzi and his collaborators of the behaviour of *Parisch. alternata* (summarized in Turillazzi 1987) indicated that intranidal relations among females were notably milder than those of *Parisch. nigricans serrei* and some species of the *Parisch. jacobsoni* group. There were rare occasions of intranidal aggression; *Polistes*-type dominance hierarchy was not seen. There was a tendency for females with mature ovaries to remain on the nest, and those with immature ovaries to perform extranidal activities, but foragers often had developed ovaries and even the female with the most developed ovaries performed extranidal tasks.

Besides the genus *Parischnogaster*, there are cases of the coexistence of potential egg-layers in a nest in other stenogastrine species. Hansell *et al.*

(1982) found that in *Liostenogaster flavolineata* subordinate females often oviposited. In *Eustenogaster calyptodoma*, the number of females attending each nest was usually one, but two or more females with developed oocytes sometimes coexisted. Young females of this species tended to leave their natal nests and found their own nests. As the proportion of failed nests was often high, due to the poor defensive capacity of single-female nests, joining an alien nest could be beneficial for the joiners. Females which failed to establish their own nests often tried to usurp the nests of other females or to occupy orphan nests. As new females could emerge soon after usurpation, the usurper would enjoy the assistance of young non-kin females. As in *Parish. mellyi*, females of *E. calyptodoma* sequentially took over the position of nest-owner; this might induce a low relatedness among second-generation females. Young *E. calyptodoma* females often remained on their natal nests or returned there after failing to found a nest. Young females performed foraging more often than old females, but there was no positive correlation between female group size and the growth of nests. According to Hansell (1987), a possible reason for coexistence is the defence of nests against predators and usurpers.

Hansell (1985) stated that the social life of *E. calyptodoma* supports the polygynous family hypothesis of West-Eberhard (1978) more than the 3/4 relatedness hypothesis of Hamilton (1964) (see also Hansell 1987). This signals a notable change in Hansell's views (see p. 98).

Thus, there is a large divergence among the social lives of the subfamily Stenogastrinae. Some species found their nests by single-female-founding, others by multi-female-founding. Although typical dominance-hierarchy and, at least partial, monopolization of oviposition by α-females were seen in some species, there are other species in which multiple potential egg-layers coexist in a nest. Even in species in which each colony has a single egg-layer, subordinate (or newly emerged) females may rear non-kin larvae or larvae with low relatedness, due to the sequential alternation of egg-laying females. Reproductive division of labour is not well developed in this subfamily, however. For example, colonies of *Parisch. alternata* may be better regarded as cases of communal nesting among almost equipotent females (but for another extreme, see *Parischnogaster* sp.; Fig. 10.1).

I believe that the basic social system of many stenogastrine species is sequential pleometrosis, a possible alternative to the simultaneous pleometrosis in *Ropalidia* and other polistine genera. A few stenogastrine species practise simultaneous pleometrosis. A possible reason for the evolution of sequential pleometrosis is the heavy pressure of predation, parasitism, and intraspecific nest usurpations. As the mean relatedness among nest members is low, this social system could be formed by mutualistic aggregation, rather than by altruism.

Why did the heavy pressure of predation, parasitism, and usurpation lead to simultaneous pleometrosis in the Polistinae, and to sequential pleometrosis in the Stenogastrinae? According to Hansell (1985*a,b*, 1987), the fact that the nests of the Stenogastrinae are constructed with mixture of soil and fibres, or virtually all soil, might be a major obstacle to the construction of large nests. This may also inhibit the formation of large aggregations of foundresses. In addition, their slender bodies and their habit of hunting prey in tropical rainforests may take them quite sensitive to attack by the Vespinae. It might be a good strategy for this group to construct many small nests and co-operate in reproduction, regardless of their low relatedness.

10.3 Social life of the *Belonogaster*

The genus *Belonogaster* includes about 35 species which occur mostly in Africa, but some species occur in Arabia (Akre 1982). In the wet tropics, nests are founded throughout the year by an association of many females (Roubaud 1916), while in the temperate zone, overwintered inseminated females found nests in the spring (Keeping and Crewe 1987). Although Roubaud (1916) considered that every female on a nest was equally capable of reproduction (that is, they are quasisocial in the sense of Fig. 2.5), Richards (1969) found differences in the performance of extranidal work and in the ovarian condition among females.

The Somalian species *B. griseus* initiates its nests by multi-female-founding (Marino Piccioli and Pardi 1970). On a 9-female nest, a linear dominance hierarchy was observed, and the α-female performed 58 per cent of dominance acts and laid 77 per cent of eggs during 112 hours of observation. The dominant female ate the eggs laid by the subordinates. The frequency of dominance-aggressive acts per female per hour was 0.31. Although dominance-aggressive acts were rare on some nests (0.075 for nest III and 0.025 for nest IV), Marino Piccioli and Pardi (1970) considered that the social system of *B. griseus* is similar to that of the typically hierarchical *Polistes* species.

Keeping and Crewe (1987) made field observations on the social behaviour of a temperate zone species, *B. petiolata*, near Transvaal, South Africa. A nest of this species was initiated by a single foundress, but other females often joined the nest during the first few weeks (*P. fuscatus*-type multi-female-founding). The ratio of multi-female-founding was 53 per cent (n = 81) and the mean foundress group size was 6.7 (n = 74). Linear hierarchy was established among cofoundresses through aggressive interactions, and only the α-foundresses became egg-layers. The functional queen (α-foundress) performed 91 per cent of dominance-aggressive acts, laid 97 per cent of the eggs, and performed 98 per cent of the oophagy (during 277 hours of observation for 22 colonies). Subordinate foundresses were therefore almost totally unable to reproduce. Dissections showed that the ovaries of most subordinate foundresses had degenerated, but β-foundresses recovered their egg-laying function when the authors removed α-foundresses.

Marking experiments showed that only 15 per cent of foundress associations consisted exclusively of nest-mates of the previous autumn. Thus, cofoundresses were not necessarily closely related.

The survival rate of multi-female colonies was 42 per cent, while single-female colonies never produced pupae during three years of observations. There was a significant positive rank correlation between colony survival rate and foundress group size. The ratio of cases in which initial foundress(es) were evicted by alien females was 61 per cent in single-female colonies and 35 per cent in multi-female colonies ($P < 0.05$, $n = 64$, χ^2 test). Although the subordinate foundresses were virtually unable to produce female progeny, they have the chance to relieve the reproductive position and to enjoy inclusive fitness gain (although this is not particularly large, because the mean relatedness value is low), and a small chance of direct reproduction. These factors might favour their joining the foundress association. If subordinate foundresses can survive until summer, when the number of nest-mates becomes large and the α-foundresses cannot monopolize oviposition effectively, multiple egg-layers may coexist, as I showed in large nests of *R. variegata* and *R. revolutionalis*. This is a subject for future study.

Thus, Keeping and Crewe (1987) concluded that (1) as single foundresses are incapable of producing colonies, joining behaviour is obligatory for the survival and reproduction of foundresses, irrespective of individual reproductive capacity; (2) the low relatedness among cofoundresses and the high rate of initial foundress displacement by foreign females makes kin selection an unlikely explanation for foundress associations in this species; (3) the lack of bimodality in body size or ovarian development among pre-nesting gynes indicates that queens are not manipulating their brood into queen-like and worker-like potential foundresses, and that parental manipulation is therefore of limited importance.

I contend that the genus *Belonogaster* can adopt two types of social systems, *Polistes dominulus*-type (multi-female-founding, followed by functional haplometrosis) and *R. revolutionalis*-type (multi-female-founding, followed by functional haplometrosis and then by true pleometrosis). This again demonstrates the great flexibility of social behaviour in primitively eusocial wasps.

11

Origin of pleometrosis: altruism or mutualism?

11.1 Mutualistic co-operation under conditions in which haplometrosis is almost impossible

I noted in Chapter 4 that the existence of pleometrosis (the coexistence of multiple egg-layers) raises a difficult problem for the kin-selection hypothesis for the evolution of eusociality, because pleometrosis reduces mean relatedness among colony members. This is also the case for Alexander's parental manipulation hypothesis (Alexander 1974).

As has been noted in Chapter 4, one of the means by which this difficulty could be overcome would be if high mean relatedness was maintained among most progeny females through functional (secondary) haplometrosis.

The existence of such a temporal pleometrosis/functional haplometrosis has been shown in many wasp species, such as *P. dominulus* and *P. fuscatus*, but the observations I described in the preceding five chapters suggest the common existence of a species in which intranidal dominance relations are mild, such that many colony members can leave their reproductive progeny on one nest. This situation does not seem to be rare in the wet tropics.

Insect populations in the wet tropics suffer strong pressure from predation, in particular from ants, the numbers of species and individuals of which are very much larger in the tropics than in temperate zones (see Jeanne 1979b for the role of predators in *Polistes* populations of temperate and tropical areas). Army ants (*Eciton* spp.) in the New World tropics and driver ants (*Dorylus* spp.) in the Old World tropics are notably specialized for hunting social insects (social wasps and other ants are attacked by *Eciton* and termites by *Dorylus*). Members of the Vespinae are the most important predators of social wasps in the Asian tropics. In addition, mature colonies of ants, bees, and wasps are very nutritious food sources for birds and mammals, if they can tolerate being stung. There are many specialized social insect predators, e.g. anteaters and pangolins (which prey on ants and termites, respectively), and honey buzzard (which prey on wasps). In tropical areas with very high species diversity and a high density of eusocial wasps, intra- and inter-specific attacks on nests are important mortality factors among wasps (*P. canadensis* is the major natural enemy of other eusocial wasps in Panama; Itô, personal observations).

Strong winds are rare in tropical rainforest areas. Conversely, tropical oceanic islands lie within regions in which typhoons or hurricanes are common. Although the predation pressure on wasp colonies on these islands may be lower than that in inland areas, adaptations to avoid the total destruction of colonies by typhoons or hurricanes are indispensable for the survival of such eusocial wasps. I showed in Fig. 6.1 that the mean productivity of cells per foundress in Okinawan *Ropalidia fasciata* was higher in pleometrotic colonies than in haplometrotic ones. Single-female colonies did not usually reconstruct their nests after destruction of their original nests by typhoons, while multi-female colonies did rebuild their nests (Table 6.1, p. 36; Iwahashi *et al.*, unpublished).

Pleometrosis inevitably involves some cost for each foundress; if a foundress is smaller than her counterpart, her ovarian development may be suppressed, or eggs laid by her will be eaten by the larger foundresses, and, even when the foundresses are equipotent, competition among cofoundresses may reduce the individual fitness of each member. Even for the larger females, performing dominance-aggressive acts may result in energetic costs. Thus, without the high selective pressures due to predation or strong winds, haplometrosis might be a better strategy than pleometrosis for individual reproduction.

Under strong external pressures, however, the situation is different. A haplometrotic nest is not guarded when its single foundress leaves the nest for foraging, but on a multi-female nest at least one of the cofoundresses can defend the nest. Multi-female colonies usually produce progeny (workers) which provide a workforce for the defence of colonies at an earlier stage than the single-female colonies (in this respect, at least in the earlier stage of colony cycle, it is often more advantageous for the dominant individual to reduce or scale down the dominance behaviour). It must be noted that all the single-female colonies of the Panamanian species studied failed to survive (although the sample size is very small) (p. 28 Table 5.2).

Assume a habitat condition in which single-female-founding is almost impossible, i.e. fitness of a genotype which leads the bearer to found a nest alone is nearly zero. Under such a condition, as argued by Lin and Michener (1972), a genetic trait for communal nest foundation may be favoured and may ultimately spread throughout the population (Hamilton 1972 also recognizes this possibility). If the founding association consists of sisters, the trait may increase much more rapidly than non-kin associations. But if (1) the rate of progeny female production is low, and (2) due to absence of winter, which synchronizes wasp colony cycles in temperate areas, progeny females emerge at a low rate throughout the year, and thus it is rather difficult for them to found a nest by association with large numbers of sisters, then, founding associations of distant relatives and even non-kin associations may provide a higher fitness for each association member than would single-female founding. For this to occur, however, two conditions must be met: (1) every foundress which joins the founding association must be able to leave at least some reproductive progeny, and/or (2) mean longevity of the dominant foundress must be short and a subordinate must be able to take over the position of the dominant foundress. For condition (2), Hamilton (1964) argued that, if the probability that a foundress belonging to a group of *n* foundresses survives until the production of reproductive progeny is more than *n* times greater than the probability of survival of a female which founds her nest alone, then temporal aggregations of non-kin females can evolve.

Compared with highly productive eusocial species, such as *Polybia*, *Vespa*, or honey-bees, the productivity of reproductive progeny in colonies of *Ropalidia* and *Mischocyttarus* is low; usually less than 200.[11]

Even when we assume that, in these two genera, all of the progeny females produced can be inseminated, the number of progeny females that emerge per nest may be less than 30 per month. This condition favours multi-female-foundation by lowly related females.

Besides swarm-founding multi-female species (see Chapter 9), all the multi-female-founding wasp species so far studied show at least partial division of reproductive tasks among foundresses; both queen-like and worker-like foundresses. Thus, there might be large variation in their fecundity. However, if condition (1) is maintained, that is, the lifetime reproductive success of even subordinate females is greater than that of females which found their nests alone, even small (subordinate-to-be) females may join the aggregation. In this case, a trait which leads the dominant female to attack subordinates seriously, or to completely monopolize the production of reproductives, may be selected against.

Thus, associative colony foundation could have evolved in the wet tropics through Lin and Michener's (1972) mutualistic aggregation hypothesis, rather than through kin-selection-based 'altruism'.[12]

---

[11] Jeanne (1980) considered that these tropical polistine wasps evolved a trait to produce a small number of less-fecund reproductives in their short colony cycle, rather than producing greatly fecund reproductives, as honey-bees and stingless bees do during their long colony cycle, and that the former may be more adaptive in tropical forests.

[12] A remaining problem is why only *Polistes canadensis* exhibited severe intranidal aggressive behaviour among the four Panamanian wasp species I studied. I cannot answer this question at present. Some points mentioned in section 6.1 must be noted, however. *P. canadensis* is the most common paper wasp found around the Panama Canal. Its body size is the largest among all paper wasps, except for some rare species, and it makes very large nests. Its females are quite aggressive to humans, which suggest that birds and small mammals would rarely attack their nests. *P. canadensis* is the major predator of other social wasps in Panama (*P. versicolor* and two *Mischocyttarus* species; Itô, unpublished). I believe that intranidal aggressiveness and the special submission posture of *P. canadensis* (Fig. 7.1) may be related to the above-mentioned life-style. In

11.2 A model of mutualistic aggregation

If f_n is the individual fitness of a foundress living in a group of n foundresses without any dominance hierarchy, f_s is the individual fitness of a solitary foundress, r is relatedness (which is assumed to be the same within group members and between a group member and a solitary one), and m is the number of solitary females which can join the group, then fitness is thought to change with the number of cofoundresses according to the curve shown in Fig. 11.1 (compare with Fig. 6.1). In this figure, n_0 is the foundress group size giving the highest fitness value. If we do not consider the inclusive fitness effect, we can expect that: (1) solitary females will wish to join the group until the group size reaches n_1 (see Fig. 11.1); and (2) females in the group will welcome a new joiner until the group size reaches n_0, but will not welcome it above this size. If we incorporate the inclusive fitness effect, however, the situation is much more complex. The mean inclusive fitness of cofoundresses, F_g, in a group of n cofoundresses is:

$$F_g = f_n + r(n-1)f_n + rmf_s \qquad (11.1)$$

Here m is the number of solitary foundresses. The inclusive fitness of a solitary foundress is

$$F_s = f_s + rnf_n + r(m-1)f_s \qquad (11.2)$$

If a solitary foundress joins the group, then

$$F_g' = f_{n+1} + rnf_{n+1} + r(m-1)f_s \qquad (11.3)$$

Thus,

$$\Delta F_g = F_g' - F_g = [f_{n+1} - f_n] + [rnf_{n+1} - r(n-1)f_n] - rf_s \qquad (11.4)$$

$$\Delta F_s = [f_{n+1} + rnf_{n+1}] - [f_s + rnf_n] \qquad (11.5)$$

where ΔF_s is the change in the fitness of a solitary foundress caused by her joining the group.

From Fig. 11.1, when $1 < n < n_0$,

$$f_{n+1} > f_n.$$

And if $f_s < f_n$,

$$\Delta F_g > 0; \Delta F_s > 0.$$

Here, joining the group is beneficial for both group foundresses and solitary foundress, regardless of r. There is no contradiction between the two.

---

addition, although intranidal aggressions were frequent, many females performed abdonem vibration (this act was performed by only the top-ranked females in other species; see pp. 54–6), and had developed eggs. Thus *P. canadensis* colonies are not strictly functionally haplometrotic.

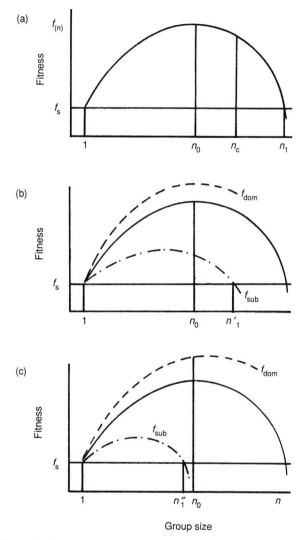

Fig. 11.1 Change in fitness with increase in the size of foundress groups. f_s, fitness of a solitary foundress. When the group size is less than n_c (or n_0 when relatedness among cofoundresses r is 0), foundresses which have already joined the group may prefer to welcome new attendants, while at group sizes in the range n_c to n_1, the foundresses may prefer to reject newcomers. Conversely, the solitary foundress may prefer to join the group until n_1 is reached, provided that no dominance-hierarchy exists among cofoundresses (a). If the dominant female can lay more eggs at the expense of the subordinate, the relationships shown in (b) and (c) should be realized. If the dominant female reduces the fitness of the subordinates too much (c), the dominant female may prefer to welcome subordinates (group size $<n_0$) while the subordinates may prefer to leave the colony ($f_{sub} < f_s$). (After Itô 1986*b*.)

If $n=n_0$ and $f_{n+1}=f_n=F$, then:

$$\Delta F_g = rnF - r(n-1)F - rf_s = r(F-f_s) \tag{11.6}$$

$$\Delta F_s = F - f_s \tag{11.7}$$

As $F>f_s$, from Fig. 12.1,

$$\Delta F_g = 0, \quad \text{when } r=0$$
$$\Delta F_g > 0, \quad \text{when } r>0$$
$$\Delta F_g = \Delta F_s, \quad \text{when } r=1.$$

If $F \geq f_s$, then $\Delta F_s > 0$. Thus, joining the group is again beneficial for both, regardless of r. When $F<f_s$, however, both ΔF_g and ΔF_s become less than 0, providing that joining the group is unfavourable for both the group and the solitary foundresses. Here, again, there is no contradiction.

If $n>n_0$, that is $f_n>f_{n+1}$, from eqn (11.4),

$$\Delta F_g = -\Delta + r[f_{n+1} - (n-1)\Delta] - rf_s \tag{11.8}$$

where $\Delta = f_n - f_{n+1}$.
Here, if $r=1$,

$$\Delta F_g = f_{n+1} - n\Delta - f_s \tag{11.9}$$

when $f_{n+1}-f_s=n\Delta$, $\Delta F_g=0$. This is point n_c in Fig. 11.1. Here $\Delta F_s = f_{n+1} - n\Delta - f_s$.
Thus, there is no contradiction between group and solitary foundresses.
Conversely, if $r=0$,

$$\Delta F_g = -\Delta < 0 \tag{11.10}$$

and $n_c = n_0$.
From eqn (11.5),

$$\Delta F_s = f_{n+1} - f_s. \tag{11.11}$$

As ΔF_g becomes negative, due to the joining of a solitary foundress, foundresses which have already joined a group may prefer to reject a new joiner while, as ΔF_s should be larger than 0 until f_{n+1} becomes equal to f_s, the solitary foundress may prefer to join the group; there is contradiction between the two. In other words, when $r=0$, group foundresses begin to reject the newcomer at a group size n_c, while the solitary foundress may prefer to join the group until the group size reaches n_1.

If there is dominance-hierarchy among cofoundresses, then the fitness curve for the dominant (f_{dom} in Fig. 11.1b and c) will be different from the curve for subordinates (f_{sub}). If the dominance hierarchy is too severe, so that f_{sub} becomes less than f_s at a point n''_1 which is less than n_0, then the subordinates may prefer to leave the group at a group size at which the dominant prefers to accept joiners. In such a situation, a trait which induces severe dominance-hierarchy may be selected against; and mild relations among cofoundresses, such as occurs in *P. versicolor* and two *Mischocyttarus* species (pre-emergence stage), may be expected.

Vehrencamp (1983, see also Vehrencamp 1984) presented a similar, inclusive fitness model, and discussed the conditions required for subordinate birds to remain in a communal breeding colony.

11.3   Eusocial insects other than wasps

Halictine and xylocopine bees

Pleometrosis among eusocial bees has been little studied until recently. For the subfamily Halictinae (Halictidae), which contains semisocial and rudimentary eusocial species, however, the number of papers reporting multi-female-founding and pleometrosis is rapidly increasing.

Packer (1986) studied the difference in fitness between dominant and subordinate females in pleometrotic colonies of *Halictus ligatus*. Near Toronto, Canada, 14.4 per cent of colonies of this species were founded by foundress association and the maximum number of foundresses per nest was four. Many of the cofoundresses were thought to be sisters or cousins which had emerged from a single nest the previous autumn (but no electrophoretic data are available to confirm this). Excavation of 13 pleometrotic nests showed that subordinate foundresses were signficantly smaller than both dominant foundresses in multi-foundress nests and those foundresses that nested solitarily. The dominant foundresses were the largest females. In this study, dominant individuals were defined as those females in multi-foundress nests that always remained on the nest during the spring foraging period. The subordinates in such nests were, surprisingly, smaller than the progeny (workers) ($P < 0.001$). The productivity of the first brood was as follows:

1. One-female colonies—6.65/♀
2. Two-female colonies—12.2 (6.2/♀)
3. Three-female colonies—37 (12.3/♀)
4. Four-female colonies—21 (5.25/♀).

The difference between the productivities per female was not significant, but multi-foundress colonies produced significantly larger numbers of reproductives (32.5, n = 2) than single-foundress colonies (15.2, n = 5). Nine of 20 single-foundress colonies failed to produce a worker brood, and one such nest produced workers but no reproductives.

Although the difference was not statistically significant, all of the 13 multi-foundress nests produced reproductive broods. Smaller foundresses were less likely to raise a worker brood than their larger counterparts. Only one of five solitarily nesting females that were similar in size to small subordinates succeeded in raising a worker brood, while 65 of 83 large foundresses succeeded in raising a brood ($P = 0.012$, Fisher exact probability test).

Packer (1986) estimated that subordinates laid 17 per cent of the worker brood eggs, but he also wrote that the subordinates were unlikely to make a substantial direct contribution to the reproductive brood. Assuming that cofoundresses are sisters, subordinate behaviour may be selected for if the productivity of two-foundress colonies exceeds 7/3 times that of single-foundress ones. The observed reproductive brood productivity of two-foundress colonies was double that of single-foundress colonies; not quite enough to select for subordinate behaviour. However, as small 'subordinate-to-be' females can produce only 1/4 of reproductives, compared with large females when they nest solitarily, the inclusive fitness of subordinates more than compensates for the reduced direct fitness in foundress associations when relatedness is greater than 1/4. Why then do 85 per cent of females found their colonies solitarily? A reason for this, according to Packer, is the low chance of co-operative foundation with sisters, because each female excavates a hibernaculum near the bottom of the natal nest but the hibernaculae are often widely dispersed. In addition, it is known that in this species and another Halictinae (*Lasioglossum lineare*, Knerer 1983) a worker population which is too large causes the dominant foundresses to lose their reproductive dominance.

The tribe Ceratinini of the subfamily Xylocopinae (Anthrophoridae) is another bee taxon which shows rudimentary eusociality similar to the Halictinae. There is, however, an important difference between the two: halictine bees rear their larvae by mass-provisioning of food, whereas xylocopine bees rear theirs by progressive-provisioning.

Sakagami and Maeta (1985) found, in *Ceratina japonica*, which utilizes dead plant stems with a soft pithy centre, that two overwintered females often used the same nest jointly. The females of this species often reutilized old nests, as well as initiating new nests in spring. More than 90 per cent of the new nests were initiated by solitary females, while 36 per cent of reutilized nests were used by multiple females. Dissection of females from multi-female nests showed that 53 per cent of nests ($n = 53$) had two potential egg-layers and 47 per cent had a single egg-layer. All of the inseminated females had developed ovaries. Sakagami and Maeta (1985) did not observe the behaviour of each female in the multi-female nests in the field but, using experimentally induced multi-female nests in a net cage, they observed that, in nests in which two females oviposited, larger females did not leave the nest, but performed nest defence, while smaller females performed foraging. Interestingly, the eggs which survived oophagy were mostly those laid by the smaller females.

An Australian allodapine bee, *Exoneura bicolor*, nests mainly in the dead fronds of tree ferns. Most nests of this species were found to contain multiple females, regardless of whether they were reused old nests (up to 18 females per nest) or newly founded ones (up to 6 females per nest) (Schwarz 1986). Oviposition began during winter in the old nests. At this time, each nest contained one or two inseminated egg-layers and several non-inseminated females with undeveloped ovaries. Interestingly, ovipositing females exhibited wing wear, suggesting that they were performing extranidal work, while none of the females with developed ovaries showed wing wear. This situation is similar to that seen in *Ceratina japonica*. In the old nests, the first brood females eclose in spring, mate, and begin to lay eggs; thus the colonies become pleometrotic. The relatedness among either overwintering nest-mates or cofoundresses was found to be about 0.5, suggesting that the females which had emerged from the same nests (sisters or cousins) joined the colony at nest initiation.

These facts suggest that females of *E. bicolor* change their reproductive strategy depending on whether they have mated (and become an egg-layer) or have not mated (and become a nest-defender). The latter may gain inclusive fitness. Schwarz (1986) suggested that strong predation pressure by ants and the high risk of brood parasitism (by cuckoo bees, *Inquilina*) might lead to multi-female founding, and that nest-mate recognition might have developed to avoid intraspecific brood parasitism.

Ants

In discussions of the origin of multi-queen social systems among ants, we must consider at least three differences between the social structures of ants and those of wasps and bees. (1) Typically, ant queens practise 'claustral founding', in which founding queens do not forage, but feed their young on their own degenerating wing muscles. Thus, before the emergence of the first worker, each ant colony is a closed system; even subordinate foundresses in pleometrotic colonies, if any, never go foraging (in ants, females which have emerged as alate adults and lost their wings after mating, or females which are apparently apterous but have small wing-buds are queens; they are distinctly different from workers which have no trace of wings or wing-buds). (2) In ants, queen-to-be female progeny of a single queen often do not perform mating flights but mate with their brothers within their nests, and become dealated, auxiliary queens. (3) In many species, even alate females which have mated outside often return to their natal nests and become auxiliary queens. Multi-queen systems originating from (2) or (3) are said to be secondarily pleometrotic. Thus, multi-queen colonies in ants are not necessarily founded by foundress associations. There are, however, cases of true multi-female founding (see Rissing and Pollock 1988), and the coexistence of multiple queens. There are, as in the swarm-founding wasps, problems concerning the evolution of eusociality in ants.

In *Lasius flavus* (Waloff 1957) and *Solenopsis saevissima* (e.g. Wilson 1966), the survival rate and productivity of reproductive progeny of group-founding queens were found to be higher than those of solitarily-founding queens. But in *Lasius flavus* all the queens except a dominant either emigrated or were expelled from their nests after the emergence of workers (Waloff 1957). After the emergence of workers, a *S. saevissima* female fought with another until the death of one of them (Wilson 1966). Thus, Hölldobler and Wilson (1977) reported that the colonies of most ants are haplometrotic and are founded basically by solitary queens.

Is this proposal true? Although I agree that many ant species are haplometrotic, or functionally haplometrotic, the proportion of ant species or ant colonies with multiple queens or egg-laying workers is much higher than would be expected. I will now review some examples.

In *Vermessor pergandii*, which lives in the semi-desert of Arizona, USA, many queen-to-be dealated females congregate under a stone and found a colony. Rissing and Pollock (1986) observed the behaviour of females in two types of experimental groups; type A groups consisted of members of the same foundress group, and type B groups consisted of members of different foundress group (possibly non-relatives). During the claustral (pre-emergence) period, no differences in behavioural interactions among members of type A and B groups, nor any dominance-aggressive acts were observed. Although there were differences in the intensity of intranidal work (nursing, etc.) among females, every queen worked, oviposited, and cared for eggs (there was no significant difference between any of the 15 experimental colonies). Oophagy was virtually absent.

Rissing and Pollock (1987) then described the behaviour of *V. pergandii* queens after the emergence of workers. After worker emergence, the mortality of queens rose suddenly, and fights between queens began. Injured queens were killed or taken out of the nest by the workers. At the same time, raiding by workers of other nests took place. Pleometrotic colonies produced more workers and the exits of their nests were opened earlier than in haplometrotic colonies. They began brood raiding earlier, and the targets were usually haplometrotic nests which were still in the pre-emergence stage. When Rissing and Pollock confined pleometrotic and haplometrotic colonies together in a single container, the pleometrotic colonies survived in 16 of 19 experiments. Thus, pleometrosis in this species may be favourable for a group monopolizing a limited nesting site and to achieve 'working power'. As ant societies are closed during the claustral stage, there may be no need for dominance-aggressive acts during this stage. Rissing and Pollock considered that this fact might be the major reason why there was no aggression among cofoundresses.

Why do *V. pergandii* females join the colony foundation despite the risk of being taken over by a non-kin female? Even a queen which was expelled from her nest by raiders could sometimes raid another colony and become a single egg-layer. In this species, according to Rissing and Pollock (1987, 1988), queens may join the foundation group, not because of kin-selection, but because of the 1/n probability of remaining in the colony and becoming the single egg-layer.

Until recently, all the reports on ants indicated that, even in species in which cofounding queens coexisted peacefully during their claustral stage, all but one queen was killed or expelled after worker emergence, either by the dominant queen (as in *V. pergandii*) or by the workers (as in *Myrmecocyctus minicus*; Bartz and Hölldobler 1982, and *Solenopsis invicta*; Tschinkel and Howard 1983). Rissing *et al.* (1989) reported, however, that cofoundress queens of a myrmicine ant, *Acromyrmex versicolor*, coexisted without any aggression before and after worker emergence (this is true for *Leptothorax congruus*, (Plate 25), K. Hamaguchi, unpublished). *A. versicolor* is unique, in that

Plate 25 Peaceful coexistence of queens (arrows indicate marked individuals) in a pleometrotic colony of *Leptothorax congruus*. (Photograph taken by K. Hamaguchi).

foundresses perform foraging activities before the emergence of workers. Rissing *et al.* (1989) observed that only a single foundress in each colony performed foraging and there was no relation between the foraging status and size or the ovarian condition of the queen. The within-colony relatedness ($r = -0.12$) was not significantly different from that between randomly selected queens ($r = 0$). In this species, larger colonies raided smaller colonies to collect a worker brood. Thus, in experimental arenas, only the largest colony, which was initiated by the largest number of cofoundresses, could survive. This situation is the same as that of *V. pergandii*. The reason for the difference (whether temporal or permanent pleometrosis) between the two savannah-inhabiting ant species is not yet known.

The above-mentioned cases describe the coexistence of true queens (which once had wings). Another type of coexistence of multiple egg-layers in ants is reproduction by the workers.

According to Ward (1983*a*,*b*), both *Rhytidoponera confusa* and *R. chalyhaea* have two types of colonies: in type A a single queen reproduces, while in type B multiple inseminated workers reproduce. The type A colonies produce a large number of new queens which initiate their nests solitarily, whereas colony fission is the major method of reproduction in type B. Relatedness among workers was 0.7 (close to the value of 0.75 for full sisters) in type A and 0.3 in type B. Dissection suggested that all or most of the inseminated workers were able to reproduce.

What is the reason for the maintenance of type B colonies? In the genus *Rhytidoponera*, six species living in rainforests and wet sclerophyll reproduce using queens, while queens are unknown in at least 40 species living in arid zones (Crozier and Pamilo 1986). Within the same species, type A and B colonies were known, but the proportion of type B colonies increased in those populations in arid areas. Australian arid eucalypt forests are very often burned by forest fires, and, paradoxically, are also often flooded (Australia is a country where the weather changes most dramatically and unpredictably). Ward (1983*b*) considered that the coexistence of multiple egg-layers might be far more beneficial in such an area than a single-queen social system, in which the loss of a queen inevitably results in colony extinction.

Rhytidoponera sp. 12 is another queenless species living in the arid zone. Crozier and Pamilo (1986) evaluated relatedness among workers using electrophoresis. The within-nest relatedness was about 0.16 (calculated from their results by Itô), significantly different from 0, but lower than that among cousins in haplodiploid animals. One colony had 23 mated individuals. These facts again show the possibly permanent coexistence of multiple egg-layers.

In an African queenless ant, *Ophthalmopone berthondi*, all the workers are capable of being functional reproductives (Peeters and Crew 1985). Only those that are sexually attractive during the period of male activity, however, can mate and lay female eggs. The percentage of mated laying workers (gamergates in Peeters and Crew's terminology) ranged from 2 to more than 60 and there was no aggression among them. This is thus a permanently pleometrotic species.

The most extreme case of pleometrosis was found in a Japanese ant, *Pristomyrmex pungens* (Tsuji 1988, 1990). This species does not have a queen, and all the workers lay worker eggs by thelytoky when they are young. The division of labour depends exclusively on age-polyethism; earlier egg-laying-intranidal stage and later foraging stage. As this ant lives in quite disturbed habitats and frequently moves its nests, the single-queen system might lead to a high risk of queen loss. How does this species maintain a social life without a queen? Tsuji (1990) suggested that the coexistence of workers, each of which had definite durations of egg-laying and foraging stages, might be favoured through a kind of group selection.

In short, although ants are said to be principally haplometrotic, truly pleometrotic species are not uncommon (we omit here polydomous *Formica* species). This fact again forces us to reconsider the role of mutualistic aggregation as proposed by Lin and Michener (1972).

Termites

Thorne (1982*a*, 1984) discovered that multi-queen colonies were common in a Panamanian termite, *Nasutitermes corniger*. Table 11.1 shows that 25 of 72 colonies (34.7 per cent) had multiple queens. One colony had 22 queens and 8 kings, and another had 33 queens and 17 kings.

Figure 11.2 shows the relationship between nest size and the number of immature individuals per nest, which is an index of colony productivity, for haplometrotic and pleometrotic colonies (the regression line for haplometrotic colonies shows an increase in queen productivity with age). Figure 11.2 indicates that pleometrotic colonies produce far larger numbers of workers than haplometrotic ones. Although the productivity per queen of pleometrotic colonies is almost the same as that of haplometrotic ones, the larger worker force produced by the former is thought to be beneficial for colony defence. Thorne did not observe any aggression among queens. Eggs laid by every queen were transported by workers to the same site in the nest and reared together.

Thorne (1984) released individuals from other colonies at the entrance of a nest. Soldiers and workers were killed immediately, but queens and kings were accepted into the nest, and eggs laid by the alien queens were transported to the same site as the eggs of the original queens.

As pleometrosis lowers intranidal relatedness, Thorne (1982*b*, 1984) proposed that the major evolutionary force for pleometrosis in *N. corniger* might be mutualistic co-operation (individual selection), rather than kin-selection.

Despite this idea, however, Thorne (1982*b*, 1984) also suggested that the essential qualities of termite pleometrosis might be different from that of the wasps, based on studies of wasp sociality made before 1980.

In the termites and ants, only reproductive castes eclose as alates, workers never have wings, and the former lose their wings after mating. Thus, a

Table 11.1 (a) Numbers of kings found in *Nasutitermes corniger* colonies with single and multiple queens. The colony without a king does not necessarily represent a functioning colony without a king; rather it may be that a small male escaped detection. (After Thorne 1984).

		Number of queens											
		1	2	3	4	9	10	11	14	18	19	22	33
Number of kings	0	11	3	1									
	1	33	8	2	1		1	1			1		
	2	1*							1	1			
	3	1*											
	4	1*				1							
	5			1									
	8											1	
	13	1*											
	17												1

(b) Distribution of the number of queens in nests of *Nasutitermes princeps*. (From Roisin 1987).

	Number of queens														
	1	2	3	4	5	6	7	8	10	11	12	16	17	18	>20
Number of nests	25	1	2	2	1	2	3	1	2	1	1	1	3	1	20

* New nests.

co-founding queen which has lost her wings is hardly able to initiate a new nest, even when she is dominated by another queen in her original nest. The queens of termites and ants are so specialized that the dominant ones cannot force others to perform foraging duties. There are large differences between termites/ants and wasps/bees, however, it may be important to note that the first case of permanent pleometrosis in termites was discovered in the wet tropics (see Plate 26).

Other cases of permanent pleometrosis in termites were found in another wet tropical area, Papua New Guinea (Roisin and Pasteels 1985; but for other possible cases, see Roisin 1987). Among three arboreal nasutitermitine species, *N. novarumhebridarum* was almost always haplometrotic, but most colonies of *N. polygynus* (14 out of 15) and 62 per cent of colonies of *N. princeps* had multiple queens (see Table 11.1; note that I use the terms pleometrosis and haplometrosis differently from Roisin). The maximum number of queens per colony was 105 in *N. polygynus* and 238 in *N. princeps*. Such multi-queen colonies also contain multiple kings, and the kings cluster in a thin-walled royal nodule. Roisin and Pasteels (1985) found queen dimorphism in

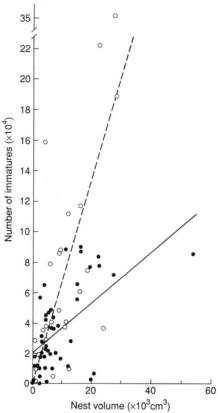

Fig. 11.2 Relation between nest size and number of immatures in *Nasutitermes corniger* in Panama. The solid and broken lines are regression lines for single-queen nests (solid circles) and multi-queen nests (open circles), respectively. (After Thorne 1984).

N. princeps; large macropterous queens and small brachypterous queens (this was the first discovery of imaginal polymorphism in the termites).

Within pleometrotic colonies all the queens exhibit the same level of physogastry whatever their morph; thus it is possible that all the queens of both types may reproduce. In *N. princeps*, the nests of pleometrotic colonies grow faster and reach a greater size than those of the haplometrotic ones. Based on data on the variation in queen numbers and percentages of pleometrotic colonies, Roisin and Pasteels (1985) considered that there may be two modes of colony reproduction in *N. princeps*: (1) a long-range reproduction by swarming, leading to haplometrotic colonies, and (2) a short-range reproduction by budding, leading to pleometrotic colonies. With models in which the starting point is the same as mine (see equations on pp. 109–11), Roisin (1987) evaluated the conditions for evolution of pleometrosis

Plate 26 A close-up photograph of part of a multi-queen nest of *Termes*? sp. (photograph taken by T. Matsumoto in West Malaysia). Workers are seen around the five queens. Soldiers are indicated by arrows.

in termites, demonstrating that pleometrosis through differentiation of queens in their natal colony (secondary pleometrosis) is easier than pleometrosis through multi-female foundation. This is consistent with the above information.

11.4   Conclusion

The foregoing accounts of bees, ants, and termites suggest that pleometrosis or multi-queen social systems could evolve in these groups under conditions of high predation pressure or high risk of colony failure. The most typical habitats in which such conditions are realized are wet tropical ecosystems. This situation is quite similar to that for the occurrence of pleometrosis in the vespine wasps.

It has been repeatedly argued that the properties of multi-queen colonies of termites (Thorne 1982*b*) and ants (Rissing and Pollock 1988; Strassmann 1989*b*) are different from those of wasps. In *Nasutitermes corniger* (Thorne 1982*a*, 1984) and *Acromyrmex versicolor* (Rissing *et al.* 1989), non-kin queens could join the nest foundation and all, or most, of them could leave reproductive progeny even after worker emergence. Even in ant species, in which all but one queen are killed or expelled by the dominant queen or workers, no aggression is observed before the emergence of workers. Conversely, according to the views of Thorne and Strassmann, cofoundresses in wasps are close kin and there is frequent aggression between them, which leads to a strict dominance hierarchy, in which only the dominant cofoundress can continue to reproduce. Strassmann (1989*b*) proposed that the high relatedness among cofounding wasp females may allow monopolization of reproduction by the dominant cofoundress, because the subordinate cofoundresses can still enjoy some inclusive fitness gain; but it is difficult for dealated ant queens to aggregate with close relatives because of their mating system (swarming) and the high risk of failure of nest initiation in soil.

I believe, however, that most social relations among cofoundresses or cofounding queens in the eusocial insects lie in an intermediate position between the two extremes: complete monopolization of gyne production by dominants (functional haplometrosis) and co-operation of almost equipotent queens (permanent pleometrosis). Although the social relations within most eusocial wasp colonies so far studied are close to the former extreme (e.g. *Polistes dominulus* and *P. fuscatus*), those of *Vespa affinis* and some (possibly most) swarm-founding wasps are close to the latter extreme.

All of the following conditions: relatedness among cofoundresses (the probability that females forming aggregations are from the same natal nest), the possibility of finding a nesting site (especially for bees) and territory (for ants and termites) for newly emerged females, predation pressure, and the longevity and maximum sustainable size of colonies (whether the colony is annual or perennial), might all be responsible for the dichotomy of reproductive patterns of multi-foundress colonies to temporal or permanent pleometrosis.

12

Manipulation of progeny by mother groups: an hypothesis for the evolution of multi-queen societies

12.1 A problem: evolution of multi-queen societies in swarm-founding Polistinae

I stated in Chapters 9 and 10 that multi-queen societies may exist in most of the swarm-founding neotropical Polistinae (*Polybia* and closely-related genera), in at least some (possibly all) of the swarm-founding species of the subgenus *Icarielia* of the genus *Ropalidia*, and in some tropical Vespinae. Although their origins may in many cases be different from that of the Vespidae, the ants also have multi-queen societies, and they have at least one species in which all the females reproduce (*Pristomyrmex pungens*, Tsuji 1990). There is no hypothesis so far to explain the evolution or maintenance of such societies, in which dozens of reproductive queens, which are morphologically distinguishable from workers, coexist (although Rosengren and Pamilo 1983 discussed the environmental conditions necessary for the development of multi-queen, polydomous societies in *Formica*).

Did the multi-queen wasp species evolve from pleometrotic, independent-founding species, such as *Mischocyttarus* spp. (see p. 59) and species of the subgenus *Icariola* (genus *Ropalidia*) (see pp. 86–9), or from haplometrotic, swarm-founding species, such as *Provespa anomala* (Matsuura and Yamane 1984)? If the latter proposal is true, then multi-queen societies might evolve through secondary pleometrosis, which is commonly seen in ants. Although there is no evidence to answer this question so far, the life history patterns of closely related taxa seem to me to suggest that, at least in some cases, the first hypothesis is possible. I will attempt to develop this hypothesis in this chapter.

12.2 Intranidal dominance relations among females in some *Mischocyttarus* and *Ropalidia* after the emergence of progeny females

Intranidal relations among the foundresses of *Mischocyttarus angulatus* and *M. basimacula* in Panama were notably peaceful during the pre-emergence period (Table 5.1). A similar situation was observed in two Australian species, *Ropalidia (Icariola)* sp. nr. *variegata* and *R. (I.) revolutionalis* (pp. 66–71).

As shown in Table 12.1, *M. angulatus* and *R.* sp. nr. *variegata* showed no dominance-aggressive acts during the pre-emergence period. During the pre-emergence periods of *M. basimacula* and *R. revolutionalis*, the frequency of dominance-aggressive acts was very low and strong aggressive acts (category 2 of Chapter 5; see p. 23 for explanation) were absent.

Table 12.1 Frequency of dominance-aggressive acts in relation to stages of the colony cycle in four tropical species of the genera *Mischocyttarus* and *Ropalidia*.

		Mischocyttarus angulatus	*M. basimacula*	*Ropalidia* sp. nr. *variegata*	*R. revolutionalis*
Pre-emergence period	*n*	3(2)	3(3)	2(2)	12(4)
	W	$0^{a,b}$	0.56 ± 0.22	0	0.04 ± 0.09^{g}
	S	$0^{c,d}$	0	0	0
	Total	$0^{e,f}$	0.56 ± 0.22	0	$0.04 \pm 0.09^{h,k}$
Intermediate period	*n*	5(1)	1	2(2)	3(2)
	W	0.62 ± 0.36^{a}	0	0	$0.13 \pm 0.23^{g,i}$
	S	0.13 ± 0.09^{c}	4.00	0	0.04 ± 0.07
	Total	0.74 ± 0.43^{e}	4.00	0	$0.22 \pm 0.16^{h,j}$
Post-emergence period	*n*	3(1)	2(1)	3(2)	5(3)
	W	0.16 ± 0.14^{b}	0.09 ± 0.12	0.82 ± 0.63	1.28 ± 0.75^{i}
	S	1.74 ± 0.58^{d}	2.51 ± 0.13	0.18 ± 0.06	1.67 ± 1.52
	Total	1.90 ± 0.44^{f}	2.60 ± 0.07	1.00 ± 0.68	$2.95 \pm 2.13^{j,k}$
Hours of observation		13.3	20.8	8.25	24.75
Data source		Itô (1984*b*)	Itô (1984*b*)	Itô (1986*c*)	Itô (1987*b*)

W, weak dominance acts; S, strong aggression (see p. 23); *n*, total number of observations. The number of nests observed is given in parentheses.

Pairs of values with the same superscripts differ significantly at the 5% (a–j) or 1% (k) level (Mann–Whitney U-test).

After the emergence of progeny females, however, such peaceful social relations drastically changed to very aggressive ones. As Table 12.1 shows, the intensity of dominance-aggressive acts rose in the four species after the emergence of progeny females, and the differences in frequencies of such acts between pre-emergence and post-emergence periods were statistically significant in *M. angulatus* and *R. revolutionalis*.

On the nests of these species, where many progeny females coexisted with foundresses, I observed frequent attacks by one female on another, which were usually followed by the attacker biting the thorax or wing of the attacked female, pulling her wings if she tried to escape, and even attempting to sting her. The attacked females ran around the nest trying to escape, flew away, or, in two *Mischocyttarus* spp., sometimes hung rigid from the attacker's mandibles. The last behaviour is thought to be a ritualized act to avoid further attacks, similar to the prostration posture observed in *Polistes canadensis* (Fig. 7.1, p. 55).

Two *Mischocyttarus* species and *R.* sp. nr. *variegata* live in the tropics, but *R. revolutionalis* lives in the subtropical, seasonal environment of Brisbane, Australia. To avoid the effect of temperature on the frequency of aggressive acts in *R. revolutionalis*, I tried to compare the number of aggressive acts divided by the number of food exchanges, because the latter may represent the activity level of colonies. The results were the same as those in Table 11.1, with the differences between pre-emergence stage, intermediate stage, and post-emergence stage, again being significant.

What is the meaning of this dramatic change in aggressiveness? My hypothesis is that the escalation of aggression is a result of 'manipulation of progeny females by the foundress group' (Itô 1986b). That is, by attacking progeny females, foundresses force their progeny to work; whereby progeny females become functional workers and may bring food to the larvae of their natal nest, irrespective of their kin-relationship (although there is a possibility that each progeny tends to give more food to their younger sisters or brothers). Under this hypothesis, most of the intranidal attacks would be performed by foundresses toward their progeny females. Although evidence for this is not yet available, most of the severest attacks I have observed in *M. angulatus* and *M. basimacula* were performed by residents toward females which had just landed on the nest (possibly returning from performing extranidal work). In addition, about two-thirds of the acts of severe aggression (category 2) were performed by larger females towards smaller ones (Table 12.2).

I could not mark all of the individuals during my short stay in Panama, but most of the larger females which attacked others were marked ones. Marked females were dominants which stayed on their nest after it was disturbed (most of the other females escaped). Thus, marked individuals on these *Mischocyttarus* nests are possibly foundresses (more than one marked female performed attacks simultaneously in both species; Table 12.2). Thus, I believe that the frequent intranidal aggression observed in the two *Mischocyttarus* spp. and *Ropalidia revolutionalis* are made by foundresses towards progeny; these attacks may then force progeny adults to work for the benefit of the younger larvae (Itô 1985a).

Table 12.2 Direction of attacks on nests of *Mischocyttarus angulatus* and *M. basimacula* during the post-emergence period. The relative size of individuals (L, large; S, small) was judged visually during encounters (added to Itô 1985*a*).

Species	L→S	L→S	S→S	S→L	Unknown	Duration of observations (min)	Number of assailants
M. angulatus	25(68)	10	2	0	20	30, 42, 72	>5*
M. basimacula	12(67)	2	2	2	4	35, 30	>4

Numbers in parentheses indicate percentage of L→S attacks to all attacks (excluding 'Unknown' column).
* Minimum estimates, because not all individuals were marked.

Of course, these nests might have subordinate foundresses which were attacked by the dominant foundress or even by dominant progeny females. As seen in Table 5.2, however, nests of *M. angulatus* and *M. basimacula* can possess multiple foundresses with developed eggs in their ovaries even after the emergence of progeny, as well as during the pre-emergence stage.

R. revolutionalis poses a different problem in this respect; on each pre-emergence stage nest, only one foundress had developed eggs in her ovaries. Despite the absence of dominance-aggressive acts the nests were functionally haplometrotic. The situation changed after the emergence of progeny, that is, many females could develop their oocytes despite frequent and severe aggression. Unfortunately, we do not know whether or not the subordinate foundresses were able to develop their ovaries.

12.3 Conditions favouring manipulation of progeny by foundress groups

My hypothesis of manipulation of progeny by the mother (foundress) group requires the following conditions: (1) all or most of the foundresses are able to lay at least some eggs after the emergence of progeny (in temperate multi-female-founding *Polistes*, only the dominant founders can produce reproductives although many foundresses on a nest can produce workers, thus they can be called 'queens' (e.g. Queller *et al.* 1988)). This condition requires some subordinates to leave reproductive progeny; (2) mothers can discriminate progeny females from cofoundresses; and (3) they cannot discriminate their own daughters from the daughters of other foundresses (or do not treat them differentially).

Condition (1) may hold true in two Panamanian *Mischocyttarus* and possibly in *R. revolutionalis*. Although each of the post-emergence nests of *R.* sp. nr. *variegata* had a single inseminated egg-layer (Itô and Yamane 1992), we must also determine the number of egg-layers on larger nests. I have no data for condition (2), but it may be possible: West-Eberhard (1978) reported that queens of *Metapolybia aztecoides* can recognize workers.

However, condition (3) is likely to be wrong, because we have accumulating evidence of precise kin-recognition ability in wasps and bees (Makino 1989; Frumhoff and Baker 1988). However, it is not unreasonable to consider that a queen in swarm-founding Polistinae does not manipulate her progeny differentially from the progeny of others (despite the possibility that she may have some ability to distinguish them).

My idea is an expansion of Alexander's parental manipulation hypothesis (p. 15). In the wet tropics, males emerge throughout the year, thus progeny females always have the possibility of being inseminated. But, as predation pressure is strong in tropical habitats, progeny females prefer to remain on their natal nest (even if their reproduction is suppressed due to manipulation by their mother) than to initiate their own nest. If the fitness of foundresses which initiate their nest alone or with a few sisters is nearly zero, they will benefit by remaining on their natal nest and rearing larvae whose mean relatedness may be lower than 0.5.

As the risk of destruction of nests and death of the dominant foundress is great in the tropics, the progeny which remain on their natal nest may have a good chance to become dominant egg-layers after the death of the 'queens'. For mothers (foundresses), there is no reason to expel helpers unless the female density exceeds a certain threshold value ($>n_c$ in Fig. 11.1).

Even in Okinawa, Japan, with its subtropical climate, progeny females of *Ropalidia fasciata* which emerged in summer can be inseminated, and they sometimes found their own nests (Itô and Yamane 1985). However, most of the progeny females of Okinawan *R. fasciata* still remain on their natal nest and perform extranidal work, as is typical of workers in temperate areas. I found only one or two cases of nest initiation by progeny females among more than 300 nest initiations.

Thus, if the great risk of nest destruction by predators or typhoons has selected for true pleometrosis and led to manipulation of progeny by mothers, the eusocial wasps approach the status of the multi-queen society as seen in *Polybia* and *Icarielia*.

13

Kin-selection and multi-queen social systems: conclusion

In this book, I have repeatedly criticized pure kin-selection views and stressed the importance of parental manipulation and mutualistic co-operation (see also Itô 1987a).

The existence of completely, or almost completely, sterile workers, as seen in honey-bees, hornets, ants, termites, and soldier-producing aphids, however, seems to be inexplicable without kin-selection.

Workers which cannot leave any progeny (including male eggs), such as the soldiers of termites and aphids, are unable to gain any fitness from mutualism with unrelated colony members. Even for workers which can reproduce after the death of queens, the probability of succeeding the queen is almost zero if the colony size is more than 1000. A genetic trait which leads mothers to sterilize their first progeny by nutritional, pheromonal, or behavioural castration, and then to produce reproductives, can be selected for if the mothers can produce more reproductive progeny than mothers without this trait (parental manipulation of Alexander 1974). At least for eusocial wasps, however, the overlapping of the emergence and of the morphological structure of workers and reproductives, and the possibility of a revolt by progeny in tropical areas makes this explanation somewhat difficult. Kin-selection provides the easiest explanation for the evolution of completely sterile workers. But the 3/4 relatedness hypothesis is not always necessary, when the left-hand side of the equation $B/C > 1/r$ is very large.

In *Polistes fuscatus*, 64 per cent of colony foundation was observed to be done by associations of females which emerged from the same nest (Noonan 1981). Strassmann (1981b) reported that all of the foundress groups of *P. annularis* were pairs of females from the same nests. In these species, as only one female (functional queen) on a nest can usually leave reproductive progeny, overwintering females which emerge from the same nest are sisters. In Okinawa, Japan, *Ropalidia fasciata* females usually overwinter on natal nests, and in spring an overwintered group splits into several small groups, each of which initates a new nest (Itô *et al.* 1985). Thus, cofoundresses are usually females which have emerged from the same nest. Although these groups of *R. fasciata* can include non-sisters (see Chapter 6), their mean relatedness may not be low.

Even in an environment in which the probability of success of single-female-founding is very low, thereby permitting pleometrosis based on mutualism, a high level relatedness among cofoundresses can afford them some inclusive fitness gains. This is the 'bonus effect', assumed to operate for helpers at the nest of the acorn woodpecker (Koenig and Pitelka 1981).

Thus, regardless of the mechanisms which have initially lead to eusociality, the kin-recognition ability may be reinforced during the life of the group. As explained in Chapter 12, when mothers manipulate their progeny, the progeny in groups with a high mean relatedness can gain higher inclusive fitness than those in low relatedness groups; in the latter case, mothers cannot manipulate the progeny so effectively, while mothers of species having advanced kin-recognition ability can manipulate the progeny more efficiently. Thus, the kin-recognition ability might have been reinforced in these species. Many temperate zone *Polistes*, such as *P. dominulus* and *P. fuscatus*, have evolved temporal pleometrosis in which only one female becomes a functional queen on a nest after the emergence of the workers. High intranidal relatedness in these species may lead subordinates to remain on their nests.

In the tropics, however, nests in different colony cycles coexist throughout the year; they are often destroyed by natural enemies and are frequently reconstructed. My impression for *P. canadensis* on Barro Colorado Island, Panama, is that reconstruction is the major way of of establishing new nests. From this situation, only one more step may be necessary to evolve swarm-founding.

My hypothetical scheme for the evolution of multi-queen colonies is shown in Fig. 13.1.

In the wet tropics, the first step in the evolution of eusociality, from the nest sharing as seen in *Zethus* (West-Eberhard 1987) and *Auplopus* (Wcislo *et al.* 1988), might be the multi-foundress-independent-founding colonies, as observed in *Mischocyttarus angulatus* and *basimacula*. Such a life-style can also be found in primitive *Ropalidia* (subgenus *Icariola*) and *Belonogaster*. The Stenogastrinae may have adopted serial polygyny due to the constraints of their nest-building material (their nests are made of mud, which is generally an unsuitable material for constructing large nests), while the pulp-nesting Polistinae have proceeded to construct large nests. Because of the frequent destruction and reconstruction of nests, the number of colonies might increase mainly through nest reconstruction processes, which are effected by females belonging to two overlapping generations. There might also be some reproductive dominance-hierarchy due mainly to differences in the nutritional

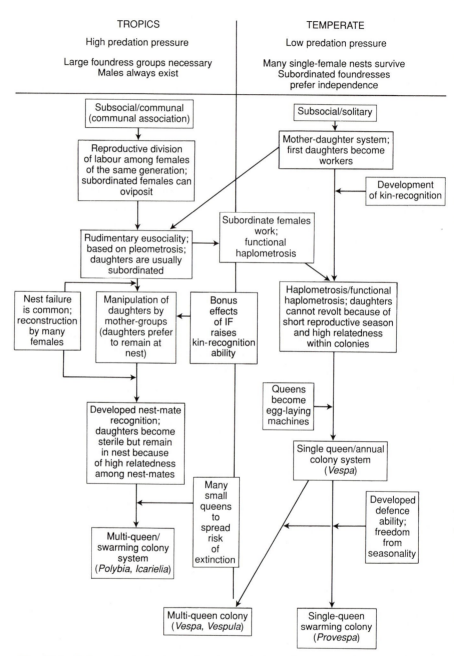

Fig. 13.1 A hypothetical scheme for the evolution of multi-queen social systems. IF: inclusive fitness.

conditions between individuals. Small individuals may therefore be produced during the early stages of nest reconstruction. Rudimentary castes would thus appear.

If manipulation of progeny females by mother groups has taken place in a colony, the inclusive fitness effect may reinforce the colony's kin-recognition ability and closeness, giving rise to a pleometrotic swarm-founding colony. Multi-queen colonies might have originated from such a colony, by the differentiation of workers and reproductives. Swarm-founding by hundreds or thousands of individuals may be one of the best strategies to overcome the great risks of predation. This, in turn, must develop the kin-recognition ability further, as seen in *Agelaia areata*, in which swarming groups follow a pheromone trail (Jeanne 1975b). This scheme is an extension of the polygynous family hypothesis proposed by West-Eberhard (1978b).

In areas where predation pressure is not very high and single-female-founding is possible, the manipulation of progeny by single mothers and kin-selection might be the major factors in caste differentiation (right-hand side of Fig. 13.1). Without predation and destruction of nests, haplometrosis may be the best way to increase fitness, because there is no reproductive competition among foundresses (in competitive situations, subordinates lose at least part of their reproductive success, and even for the dominants there must be some cost involved in dominating others). It must be mentioned here that reproductive equivalency of potential egg-layers has not been observed in any polistine multi-female-founding species, except the *Polybia-Icarielia* group.

One problem with the scheme in Fig. 13.1 is why honey-bees and many stingless bees have adopted the single-queen swarm-founding system while the *Polybia*-group and *Icarielia* have adopted multi-queen swarm-founding systems. Queens in large, single-queen colonies are 'oviposition machines'. Queens of ants, termites, and honey-bees lay more than 1000 eggs per day. Conversely, queens of the *Polybia*-group and *Icarielia* are only slightly larger (in some species they are smaller) than their workers, and can lay only a few to dozens of eggs per day. These species have evolved to have many small queens in a colony, rather than a single large queen. This strategy may be beneficial when the risk of complete destruction of nests by army ants or mammals is large, because a big oviposition machine cannot fly well and may die within the destroyed nest. Multi-queen swarm-founding systems may be sustained under these conditions, as an evolutionarily stable strategy.

Fig. 13.2 shows a tentative diagram for the evolution of eusociality in the Vespidae, in which even the most developed systems can evolve through pleometrosis. Here the *M. angulatus*-type wasps are considered to face a bifurcation in their route to highly eusocial societies. The first possibility is to develop temporary pleometrosis, as in *R. cincta*, *R. marginata*, *M. drewseni*, and *B. griseus*. But, along this route, it may be difficult to develop foundress associations with a hundred or so females. Another option is to evolve a

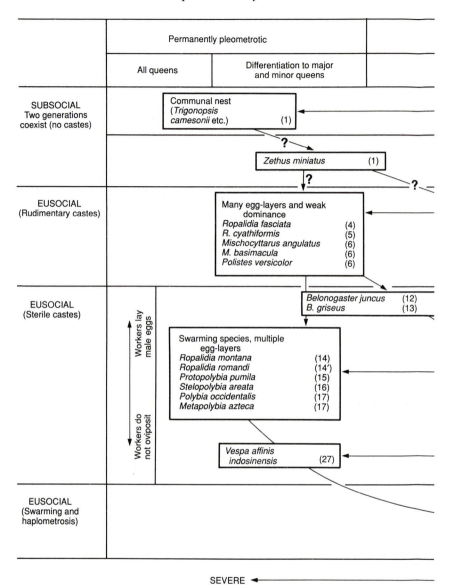

Fig. 13.2 Hypothetical pathways for the evolution of eusociality in the Vespidae. Taxon belonging to other families are shown in parentheses. Data sources: (1) West-Eberhard (1978*b*); (2) S. Yamane *et al.* (1983b); (3) Hansell (1982); (4) Itô (1983*a*,*b*); (5) Gadagkar and Joshi (1982*a*, 1984); (6) Itô (1984*b*, 1985*a*); (7) Darchen (1976*a*); (8) Gadagkar and Joshi (1982*b*, 1983); (9) Sekijima *et al.* (1980); (10) Litte (1979); (11) Jeanne (1972); (12) Roubaud (1916); (13) Marino Piccioli and Pardi (1970);

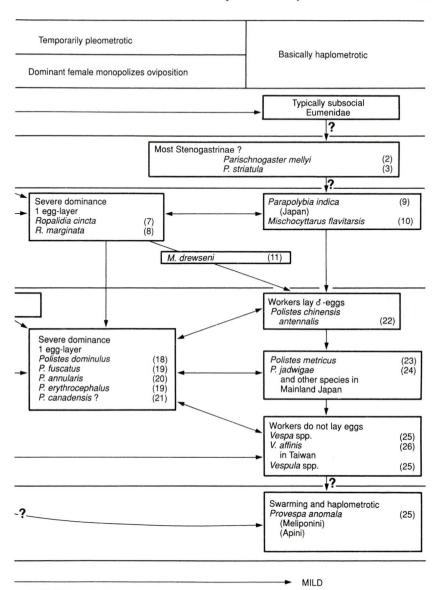

(14) Yamane *et al.* (1983*a*); (14') Shima-Machado and Yamane (unpublished); (15) Naumann (1970); (16) Jeanne (1975*a,b*); (17) Forsyth (1978); (18) Pardi (1948); (19) West-Eberhard (1969); (20) Strassmann (1983); (21) Pickering (1980); (22) Miyano (1980) and N. Miyano (personal communication); (23) Starr (1976); (24) E. Kasuya (personal communication); (25) Matsuura and Yamane (1984); (26) S. Yamane (personal communication); (27) Matsuura (1983).

multi-queen giant colony system based on a developed kin-recognition system. This, in my view, could evolve from pleometrotic independent-founding species.

Based on cladistic analyses and data on social behaviour in the Vespidae, Carpenter (1989) reached the opinion that caste formation preceding long-term monogyny (haplometrosis by my definition), and long-term polygyny (permanent pleometrosis) was always derived from monogyny and that there is no evidence for a polygynous transition, that is its evolution from a 'rudimentary-caste-containing' stage. It is true that many of the rudimentary-caste-containing species (*Polistes*, *Mischocyttarus*, and *Ropalidia* (*Icariola*)) so far studied are monogynous or functionally monogynous. But it must be noted that all polistine species which have morphologically distinct castes are pleometrotic, and at least *Ropalidia* (*Anthreneida*) *sumatorae* and *R.* (*Icariola*) *socialistica* are possibly permanently pleometrotic despite the lack of morphological castes. The possibility of 'polygynous transition' cannot be excluded.

To determine what part of this scheme can be retained and what part must be altered, we must perform long-term studies in the tropics and clarify the following important points:

1. The relative contributions of dominant and subordinate females to the production of reproductives after the emergence of progeny in individual nests of tropical, multi-female independent-founding wasps.
2. Whether the escalated intranidal aggression in *Mischocyttarus* and *Ropalidia* following the emergence of progeny is an indication of functional haplometrosis or manipulation of progeny by mother groups. For this, we must know (a) whether most attacks are made by foundresses toward progeny, and (b) whether many females can lay eggs that survive after the escalation of aggression.
3. The social structure (especially the number of egg-layers) of species such as *Ropalidia socialistica*, *R. trichophtalma*, and *R. sumatorae*, which are considered to be transitional in social structure between *Icariola* and *Icarielia*.
4. The social behaviour of multi-queen swarm-founding species.
5. The differences in kin-recognition ability among species at different stages along the hypothetical routes of social evolution.
6. The effects of nest perenniality on pleometrosis.

It must be mentioned again that the kin-selection/inclusive fitness theory has created a new era in understanding the evolution of insect sociality. However, the one-sided view of kin-selection has also set a constraint on further studies. We can generate many new study projects by liberating ourselves from this view.

References

Akre, R. D. (1982). Social wasps. In *Social insects*, Vol. 4 (ed. H. R. Hermann), pp. 1–105, Academic Press, New York.

Alexander, R. D. (1974). The evolution of social behaviour. *Annual Review of Ecology and Systematics* **5**, 325–81.

Andersson, M. (1984). The evolution of eusociality. *Annual Review of Ecology and Systematics* **15**, 165–89.

Aramaki, H. (1985). Female dimorphism in the tropical paper wasp, *Ropilidia sumatorae*. Unpublished B.Ed. thesis, Ibaraki University.

Axelrod, R. and Hamilton, W. D. (1981). The evolution of cooperation. *Science* **211**, 1390–6.

Bartz, S. H. and Hölldobler B. (1982). Colony founding in *Myrmecocystus mimicus* Wheeler (Hymenoptera: Formicidae) and the evolution of foundress associations. *Behavioral Ecology and Sociobiology* **10**, 137–47.

Belavadi, V. V. and Govindan, R. (1981). Nesting habits and behaviour of *Ropilidia* (*Icariola*) *marginata* (Hymenoptera: Vespidae) in south India. *Colemania* **1**, 95–101.

Bohm, M. F. K. (1972). Reproduction in *Polistes metricus*. Unpublished Ph.D. thesis, University of Kansas, Lawrence (cited in Starr, 1976).

Brockmann, H. J. (1984). The evolution of social behaviour in insects. In *Behavioural ecology: an evolutionary approach* (ed. J. R. Krebs and N. B. Davies, pp. 340–61. Blackwell, Oxford.

Brothers, D. J. (1974). Phylogeny and classification of the aculeate Hymenoptera, with special reference to Mutillidae. *University of Kansas Science Bulletin* **50**, 483–648.

Carpenter, J. M. (1982). The phylogenetic relationship and natural classification of the Vespoidea (Hymenoptera). *Systematic Entomology* **7**, 11–38.

Carpenter, J. M. (1989). Testing scenarios: wasp social behaviour. *Cladistics* **5**, 131–44.

Carpenter, J. M. (1991). Phylogenic relationships and the origin of social behaviour in the Vespidae. In *The social biology of wasps* (ed. K. G. Ross and R. W. Matthews), pp. 17–32. Cornell University Press, Ithaca.

Carpenter, J. M. and Day, M. C. (1988). Nomenclatural notes on Polistinae (Hymenoptera: Vespidae). *Proceedings of the Entomological Society, Washington* **90**, 323–8.

Carpenter, J. M. and Ross, K. G. (1984). Colony composition in four species of Polistinae from Suriname, with a description of the larvae of *Brachygastra scutgellaris* (Hymenoptera: Vespidae). *Psyche* **91**, 237–50.

Cervo, R. and Turillazzi, S. (1985). Associative foundation and nesting sites in *Polistes nympha*. *Naturwissenschaften* **72**, 48–9.

Crozier, R. H. and Pamilo, P. (1986). Relatedness within and between colonies of a queenless ant species of the genus *Rhytidoponera* (Hymenoptera: Formicidae). *Entomologica Generalis* **11**, 113–17.

Cruz, Y. P. (1981). A sterile defender morph in a polyembryonic hymenopterous parasite. *Nature* **294**, 446–7.

Cumber, R. A. (1951). Some observations on biology of the Australian wasp *Polistes humilis* Fabr. (Hymenoptera: Vespidae) in north Auckland (New Zealand), with special reference to the nature of the worker caste. *Proceedings of the Entomological Society London* **A26**, 11–16.

Darchen, R. (1976a). *Ropilidia cincta*, guêpe sociale de la savane de Lamto (Côte-d'Ivoire) (Hym. Vespidae). *Annales de la Société d'Entomologie Française* (*N.S.*) **12**, 579–601.

Darchen, R. (1976b). La formation d'une nouvelle colonie de *Polybioides tabius* Fab. (Vespidae, Polybiinae). *C. R. Académie des Sciences Paris* **282**, D-457–9.

Darwin, C. (1859). Origin of species by means of natural selection. In *The origin of species and the descent of man* (Charles Darwin), pp. 203–7, The Modern Library, New York.

Davis, T. A. (1966). Observations on *Ropilidia variegata* (Smith) (Hymenoptera: Vespidae). *Entomological News* **27**, 271–7.

Douglas, A. and Servenity, V. N. (1951). The establishment of the paper nest wasp in Western Australia. *Western Australian Naturalist*, **2**, 169–74.

Evans, H. E. (1953). Comparative ethology and the systematics of spider wasps. *Systematic Zoology* **2**, 155–72.

Evans, H. E. and West-Eberhard, M. J. (1970). *The wasps.* University of Michigan Press, Ann Arbor.

Fisher, R. A. (1930). *The genetical theory of natural selection.* Clarendon Press, Oxford.

Forsyth, A. B. (1978). Studies on the behavioural ecology of polygynous social wasps. Ph.D. dissertation, Harvard University, Cambridge, Massachusetts.

Frumhoff, P. C. and Baker, J. (1988). A genetic component of division of labour within honeybee colonies. *Nature* **333**, 358–61.

Gadagkar, R. (1980). Dominance hierarchy and division of labour in the social wasp, *Ropalidia marginata* (Lep.) (Hymenoptera: Vespidae). *Current Science* **49**, 772–5.

Gadagkar, R. (1987). Social structure and the determinants of queen status in the primitively eusocial wasp, *Ropilidia cyathiformis*. In *Chemistry and biology of social insects, Proceedings of the International Congress of the IUSSI, München* (ed. J. Eder and H. Rembold), pp. 377–8.

Gadagkar, R. (1990). Origin and evolution of eusociality: A perspective from studying primitively eusocial wasps. *Journal of Genetics* **69**, 113–25.

Gadagkar, R. and Gadgil, M. (1978). Observations on population ecology and sociobiology of the paper wasp *Rapalidia marginata marginata* (Lep.) (Family Vespidae). *Paper presented at the Symposium on Ecology of Animal Populations. Zoological Survey of India*, Calcutta, Oct. 1978. (Copy can be obtained from the author.)

Gadagkar, R. and Joshi, N. V. (1982a). Behaviour of the Indian social wasp *Ropalidia cyathiformis* on a nest of separate combs (Hymenoptera: Vespidae). *Journal of Zoology London* **198**, 27–37.

Gadagkar, R. and Joshi, N. V. (1982b). A comparative study of social structure in colonies of *Ropalidia*. In *The biology of social insects* (ed. M. D. Breed, C. D. Michener, and H. E. Evans, pp. 187–91. Westview Press, Boulder, Colorado.

Gadagkar, R. and Joshi, N. V. (1983). Quantitative ethology of social wasps: time-activity budgets and cast differentiation in *Ropalidia marginata* (Lep.) (Hymenoptera: Vespidae). *Animal Behaviour* **31**, 26–31.

Gadagkar, R. and Joshi, N. V. (1984). Social organization in the Indian wasp *Ropalidia cyathiformis* (Fab.) (Hymenoptera: Vespidae). *Zeitschrift für Tierpsychologie* **64**, 15–32.

Gadagkar, R. and Joshi, N. V. (1985). Colony fission in a social wasp. *Current Science* **54**, 57–62.

Gadgil, M. and Mahabal, A. (1974). Caste differentiation in the paper wasp *Ropalidia marginata* (Lep.). *Current Science* **43**, 482.

Gamboa, G. J. (1978). Intraspecific defence: advantage of social cooperation among paper wasp foundresses. *Science* **199**, 1463–5.

Gamboa, G. J. (1980). Comparative timing of brood development between multiple- and single-foundress colonies of the paper wasp, *Polistes metricus*. *Ecological Entomology* **5**, 221–5.

Gamboa, G. J. and Dew, H. E. (1981). Intracolonial communication by body oscillations in the paper wasp, *Polistes metricus*. *Insectes Sociaux* **28**, 13–26.

Gamboa, G. J., Heacock, B. D., and Wiltjer, S. L. (1978). Division of labor and subordinate longevity in foundress associations of the paper wasp, *Polistes metricus* (Hymenoptera: Vespidae). *Journal of the Kansas Entomological Society* **51**, 343–52.

Gervet, J. (1964). Le comportement d'oophagie différentielle chez *Polistes gallicus* L. (Hymen. Vesp.). *Insectes Sociaux* **11**, 343–82.

Gibo, D. L. (1978). The selective advantage of foundress associations in *Polistes fuscatus* (Hymenoptera: Vespidae): a field study of the effect of predation on productivity. *Canadian Entomologist* **110**, 519–40.

Grechka, E. O. and Kipyatkov, V. E. (1984). Sezonnuii tsikl razvitiya i kastovaya determinatsiya i obshchestvennoi ocui *Polistes gallicus* (Hymenoptera: Vespidae). II. Dinamika rosta i produktivnosti kolonii. *Zoologicheskii Zhurnal* **63**, 81–94.

Haldane, J. B. S. (1932). *The causes of evolution.* Longmans, London.

Hamilton, W. D. (1964) The genetical theory of social behaviour, I and II. *Journal of Theoretical Biology* **7**, 1–52.

Hamilton, W. D. (1971). Geometry for the selfish herd. *Journal of Theoretical Biology* **31**, 295–311.

Hamilton, W. D. (1972). Altruism and related phenomena, mainly in social insects. *Annual Review of Ecology and Systematics* **3**, 193–232.

Hamilton, W. D. (1987). Kinship, recognition, disease, and intelligence: constraints of social evolution. In *Animal societies: theories and facts* (ed. Y. Itô, J. L. Brown, and J. Kikkawa), pp. 81–102. Japanese Science Society Press, Tokyo.

Hansell, M. H. (1981). Nest construction in the subsocial wasp, *Parischnogaster mellyi* (Saussure) (Stenogastrinae, Hymenoptera). *Insectes Sociaux* **28**, 208–16.

Hansell, M. H. (1982) Colony membership in the wasps, *Parischnogaster striatula* (Stenogastrinae). *Animal Behaviour* **30**, 1258–9.

Hansell, M. H. (1983) Social behaviour and colony size in the wasp *Parischnogaster mellyi* (Saussure), Stenogastrinae (Hymenoptera: Vespidae). *Proc. Koninkl, Ned. Akad. Wetens.* **C86**, 167–78.

Hansell, M. H. (1985). The nest material of Stenogastrinae (Hymanoptera Vespidae) and its effect on the evolution of social behaviour and nest design. *Insectes Sociaux* **2**, 57–63.

Hansell, M. H. (1987). Elements of eusociality in colonies of *Eustenogaster calyptodoma* (Sakagami and Yoshikawa) (Stenogastrinae, Vespidae). *Animal Behaviour* **35**, 131–41.

Hansell, M. H., Samuel, C., and Furtado, J. I. (1982) *Liostenogaster flavolineata*: social life in the small colonies of an Asian tropical wasp. In *The biology of social insects* (ed. M. D.Breed, C. D. Michener, and H. E. Evans), pp. 192–5. Westview Press, Boulder, Colorado.

Higashi, S. (1976). Nest proliferation by budding and nest growth pattern in *Formica (Formica) yessensis* in Ishikari Shore, *Journal of the Faculty of Science, Hokkaido University, Series VI, Zoology* **20**, 359–89.

Hirose, Y. and Yamasaki, M. (1984). Foundress association in *Polistes jadwigae* Dalla Torre (Hymenoptera, Vespidae): relatedness among cofoundresses and colony productivity. *Kontyû, Tokyo* **52**, 172–4.

Höfling, J. C. and Machado, V. L. L. (1985). Anályse populacional de colônias de *Polybia ignobilis* (Haliday, 1836) (Hymenoptera, Vespidae). *Revista Brasileira Entomologia* **29**, 271–84.

Hölldobler, B. and Wilson, E. O. (1977). The number of queens: an important trait in ant evolution. *Naturwissenschaften* **64**, 8–15.

Hölldobler, B. and Wilson, E. O. (1900). *The ants.* Belknap Press of Harvard University Press, Cambridge, Massachusetts.

Hook, A. W. and Evans, H. E. (1982). Observations in the nesting behaviour of three species of *Ropalidia* Guérin-Méneville (Hymenoptera: Vespidae). *Journal of the Australian Entomological Society* **21**, 271–5.

Hoshikawa, T. (1979). Observations on the polygynous nests of *Polistes chinensis antennalis* Pérez (Hymenoptera: Vespidae) in Japan. *Kontyû, Tokyo* **47**, 239–43.

Hughes, C. R. and Strassmann, J. E. (1988). Age is more important than size in determining dominance among workers in the primitively eusocial wasp, *Polistes instabilis. Behaviour* **107**, 1–14.

Imanishi, K. (1951) *Ningen-izen no syakai (Infrahuman societies).* Iwanamishoten, Tokyo, (in Japanese).

Ito, M. (1973). Seasonal population trends and nest structure in a polydomous ant, *Formica (Formica) yessensis* Forel. *Journal of the Faculty of Science, Hokkaido University Series VI Zoology* **19**, 270–93.

Itô, Y. (1980). *Comparative ecology,* (trans. J. Kikkawa). Cambridge University Press, Cambridge.

Itô, Y. (1983*a*). A note on the social behaviour of a subtropical paper wasp, *Rapalidia fasciata* (F.), on a reconstructed nest. *Kontyû, Tokyo* **51**, 269–75.

Itô, Y. (1983*b*). Social behaviour of a subtropical paper wasp, *Ropalidia fascita* (F.): field observations during founding stage. *Journal of Ethology* **1**, 1–14.

Itô, Y. (1984*a*). Shifts of females between adjacent nests of *Polistes versicolor* (Hymenoptera: Vespidae) in Panama. *Insectes Sociaux* **31**, 103–11.

Itô, Y. (1984*b*). Social behaviour and social structure of Neotropical paper wasps, *Mischocyttarus angulatus* Richards and *M. basimacula* (Cameron). *Journal of Ethology* **2**, 17–19.

Itô, Y. (1985*a*). A comparison of intracolony aggressive behaviours among five species of polistine wasps (Hymenoptera: Vespidae). *Zeitschrift für Tierpsychologie* **68**, 152–67.

Itô, Y. (1985*b*). Social behaviour of an Australian paper wasp, *Ropalidia plebeiana,* with special reference to the process of acceptance of an alien female. *Journal of Ethology* **3**, 21–5.

Itô, Y. (1985*c*). Colony development and social structure in a subtropical paper wasp, *Ropalidia fasciata* (F.) (Hymenoptera: Vespidae). *Researches on Population Ecology* **27**, 333–49.

Itô, Y. (1986*a*). Spring behaviour of an Australian paper wasp, *Polistes humilis synoecus*: colony founding by haplometrosis and utilization of old nests. *Kontyû, Tokyo* **54**, 191–202.

Itô, Y. (1986b). On the pleometrotic route of social evolution in the Vespidae. *Monitore Zoologico Italiano* (*N.S.*) **20**, 241–62.

Itô, Y. (1986c). Observations on the social behaviour of three polistine wasps (Hymenoptera: Vespidae). *Journal of the Australian Entomological Society* **25**, 309–14.

Itô, Y. (1986d). Social behaviour of *Ropalidia fasciata* (Hymenoptera: Vespidae) females on satellite nests and on a nest with multiple combs. *Journal of Ethology* **4**, 73–80.

Itô, Y. (1986e). *Karibati no syakai-sinka*. Tôkai University Press, Tokyo. (In Japanese).

Itô, Y. (1987a). Role of pleometrosis in the evolution of eusociality in wasps. In *Animal societies: theories and facts* (ed. Y. Itô, J. L. Brown, and J. Kikkawa), pp. 17–34. Japanese Science Society Press, Tokyo.

Itô, Y. (1987b). Social behaviour of the Australian paper wasp, *Ropalidia revolutionalis* (de Saussure) (Hymenoptera: Vespidae). *Journal of Ethology* **5**, 115–24.

Itô, Y. and Higashi, S. (1987). Spring behaviour of *Ropalidia plebeiana* (Hymenoptera: Vespidae) within a huge aggregation of nests. *Applied Entomology and Zoology* **22**, 519–27.

Itô, Y. and Iwahashi, O. (1987). An analysis of foundress group size in *Ropalidia fasciata* (Hymenoptera: Vespidae) with zero-truncated distributions. *Researches on Population Ecology* **29**, 289–94.

Itô, Y. and Yamane, Sk. (1985). Early male production in a subtropical paper wasp, *Ropalidia fasciata* (Hymenoptera: Vespidae). *Insectes Sociaux* **32**, 403–10.

Itô, Y. and Yamane, S. (1992). Social behaviour of two primitively eusocial wasps, *Ropalidia* sp. nr. *variegata* and *R. gregaria gregaria* (Hymenoptera: Vespidae) in Northern Territory, Australia, with special reference to task specialization and mating inhibition. *Journal of Ethology* **10**, (in press).

Itô, Y., Iwahashi, O., Yamane, S., and Yamane, Sk. (1985). Overwintering and nest reutilization in *Ropalidia fasciata*. *Kontyû, Tokyo* **53**, 486–90.

Itô, Y., Yamane, S., and Spradbery, J. P. (1988). Population consequences of huge nesting aggregation of *Ropalidia plebeiana* (Hymenoptera: Vespidae). *Researches on Population Ecology* **30**, 279–95.

Iwahashi, O. (1989). Society of *Ropalidia fasciata*. In *Societies of Ropalidia wasps* (ed. O. Iwahashi and S. Yamane), pp. 3–209. Tôkai University Press, Tokyo, (in Japanese).

Iwata, K. (1938). Habits of some bees in Formosa. IV (*Allodape*). *Transactions of the Natural History Society of Formosa* **28**, 373–9.

Iwata, K. (1942). Comparative studies on the habits of solitary wasps. *Tenthredo* **4**, 1–146, 4 plates.

Iwata, K. (1969). On the nidification of *Ropalidia* (*Anthreneida*) *taiwana koshunensis*

Sonan in Formosa (Hymenoptera: Vespidae). *Kontyû, Tokyo* **37**, 367–72, (in Japanese with English summary).

Iwata, K. (1971). *Evolution of instinct: comparative ethology of Hymenoptera* (trans. A. Gopal, 1976). Amerind Publishing Co., New Delhi.

Jeanne, R. L. (1972). Social biology of the Neotropical wasp *Mischocyttarus drewseni*. *Bulletin of the Museum of Comparative Zoology, Harvard University* **144**, 63–150.

Jeanne, R. L. (1975*a*). The adaptiveness of social wasp nest architecture. *Quarterly Review of Biology* **50**, 267–87.

Jeanne, R. L. (1975*b*). Behaviour during swarm movement in *Stelopolybia areata* (Hymenoptera: Vespidae). *Psyche* **82**, 259–64.

Jeanne, R. L. (1979*a*). Construction and utilization of multiple combs in *Polistes canadensis* in relation to the biology of predaceous moth. *Behavioural Ecology and Sociobiology* **4**, 293–310.

Jeanne, R. L. (1979*b*). A latitudinal gradient in rates of ant predation. *Ecology* **60**, 1211–24.

Jeanne, R. L. (1980). Evolution of social behaviour in the Vespidae. *Annual Review of Entomology* **25**, 371–96.

Jeanne, R. L. (1991). The swarm-founding Polistinae. In *Social biology of wasps* (ed. K. G. Ross and R. W. Matthews), pp. 191–231. Cornell University Press, Ithaca, New York.

Jeanne, R. L. and Fagen, R. (1975). Polymorphism in *Stelopolybia areata* (Hymenoptera: Vespidae). *Psyche* **81**, 155–66.

Kasuya, E. (1981). Polygyny in the Japanese paper wasp, *Polistes jadwigae* Dalla Torre (Hymenoptera: Vespidae). *Kontyû, Tokyo* **49**, 306–13.

Kasuya, E. (1982). Factors governing the evolution of eusociality through kin-selection. *Researches on Population Ecology* **24**, 174–92.

Kasuya, E., Hibino, Y., and Itô, Y. (1980). On 'intercolonial' cannibalism in Japanese paper wasps, *Polistes chinensis antennalis* and *P. jadwigae*. *Researches on Population Ecology* **22**, 255–62.

Kawai, M. (1979). Introduction. In *Ecological and sociological studies of Gelada baboons* (ed. M. Kawai). Kodansha, Tokyo and Karger, Basel.

Keeping, M. G. and Crewe, R. M. (1983). Parasitoids, commensals and colony size in nests of *Belonogaster* (Hymenoptera: Vespidae). *Journal of the Entomological Society of South Africa* **46**, 309–23.

Keeping, M. G. and Crewe, R. M. (1987). The ontogeny and evolution of foundress associations in *Belonogaster petiolata* (Hymenoptera: Vespidae). In *Chemistry and biology of social insects* (ed. J. Eder and H. Rembold), pp. 383–4. Verlag J. Peperny, München.

Klahn, J. E. (1979). Philopatric and non-philopatric foundress associations in the social wasp *Polistes fuscatus*. *Behavioural Ecology and Sociobiology* **5**, 413–24.

Klahn, J. E. (1988). Intraspecific comb usurpation in the social wasp *Polistes fuscatus*. *Behavioural Ecology and Sociobiology* **23**, 1–8.

Knerer, G. (1983). The biology of social behavior of *Evylaeus linearis* (Schenck). *Zoologischer Anzeiger* **211**, 177–86.

Koenig, W. D. and Pitelka, F. A. (1981). Ecological factors and kin selection in the evolution of cooperative breeding in birds. In *Natural selection and social behaviour: recent research and new theory* (ed. R. D. Alexander and D. W. Tinkle), pp. 261–80. Chiron Press, New York.

Kojima, J. (1984a). Division of labour and dominance interaction among cofoundresses on pre-emergence colonies of *Ropalidia fasciata* (Hymenoptera: Vespidae). *Biological Magazine, Okinawa* **22**, 27–35.

Kojima, J. (1984b). Construction of multiple independent combs in *Ropalidia fasciata* (Hymenoptera: Vespidae). *Japanese Journal of Ecology* **34**, 233–4.

Krebs, J. R. and Davies, N. B. (1987). *An introduction to behavioural ecology.* (2nd edn). Blackwell, Oxford.

Lester, L. J. and Selander, R. K. (1981). Genetic relatedness and the social organization of *Polistes* colonies. *American Naturalist* **117**, 147–66.

Lin, N. and Michener, C. D. (1972). Evolution of eusociality in insects. *Quarterly Review of Biology* **47**, 131–59.

Litte, M. (1977). Behavioural ecology of the social wasp, *Mischocyttarus mexicanus*. *Behavioural Ecology and Sociobiology* **2**, 229–46.

Litte, M. (1979). *Mischocyttarus flavitarsis* in Arizona: social and nesting biology of a polistine wasp. *Zeitschrift für Tierpsychologie* **50**, 282–312.

Litte, M. (1981). Social biology of the polistine wasp *Mischocyttarus labiatus*: survival in a Colombian rain forest. *Smithsonian Contributions to Zoology* **327**, 1–27.

Machado, V. L. L. (1985) Analyse populacional de colônias de *Polybia* (*Myrapetra*) *paulista* (Ihering, 1896) (Hymenoptera: Vespidae). *Revista Brasileira de Zoologia* **2**, 187–201.

Makino, S. (1989). Usurpation and nest rebuilding in *Polistes riparius*: two ways to reproduce after the loss of original nest (Hymenoptera: Vespidae). *Insectes Sociaux* **36**, 116–28.

Marino Piccioli, M. T. and Pardi, L. (1970). Studi sulla biologia *Belonogaster* (Hymenoptera: Vespidae). 1. Sull'etogramma di *Belonogaster griseus* (Feb.). *Monitore Zoologico Italiano (N. S.)* Suppl. 3, **9**, 197–225.

Matsuura, M. (1977). Life of *Polistes* wasps. *Shizen* **32**, 26–36, (in Japanese).

Matsuura, M. (1983). Preliminary report on polygynous colonies of *Vespa affinis indosinensis* (Hymenoptera: Vespidae) in Sumatra. *Kontyû, Tokyo* **51**, 80–2.

Matsuura, M. and Yamane, Sk. (1984). *Comparative ethology of the Vespinae.* Hokkaido University Press, Sapporo, (English translation, 1990, Springer Verlag).

Matthews, R. W. (1968). *Microstigmus comes*: sociality in a sphecid wasp. *Science* **160**, 787–8.

Metcalf, R. A. (1980). Sex ratios, parent-offspring conflict, and local competition for males in the social wasps *Polistes metricus* and *Polistes variatus*. *American Naturalist* **116**, 642–54.

Metcalf, R. A. and Whitt, G. S. (1977*a*). Intra-nest relatedness in the social wasp *Polistes metricus*: a genetic analysis. *Behavioral Ecology and Sociobiology* **2**, 339–51.

Metcalf, R. A. and Whitt, G. S. (1977*b*). Relative inclusive fitness in the social wasp *Polistes metricus*. *Behavioral Ecology and Sociobiology* **2**, 353–60.

Michener, C. D. (1958). The evolution of social behavior in bees. *Proceedings of the Xth International Congress of Entomology*, Montreal, 1956, **2**, 444–7.

Michener, C. D. (1964). Reproductive efficiency in relation to colony size in hymenopterous societies. *Insectes Sociaux* **11**, 317–42.

Michener, C. D. (1969). Comparative social behavior of bees. *Annual Review of Entomology* **14**, 299–342.

Michener, C. D. (1974). *The social behavior of bees: a comparative study*. Belknap Press of Harvard University Press, Cambridge, Massachusetts.

Michener, C. D. (1985). From solitary to eusocial: need there be a series of intervening species? In *Experimental behavioral ecology and sociobiology. Fortschritte der zoologie, Bd. 31* (ed. B. Hölldobler and M. Lindauer), pp. 293–305. Gustav Fischer, Stuttgart.

Michener, C. D. and Brothers, (1974). Were workers of eusocial Hymenoptera initially altruistic or oppressed? *Proceedings of the National Academy of Science (USA)* **71**, 671–4.

Miyano, S. (1980). Life tables of colonies and workers in a paper wasp, *Polistes chinensis antennalis*, in central Japan (Hymenoptera: Vespidae). *Researches on Population Ecology* **22**, 69–88.

Morimoto, R. (1961*a*). On the dominance order in *Polistes* wasps. I. *Scientific Bulletin of the Faculty Agriculture of Kyushu University* **18**, 339–51, (in Japanese with English summary).

Morimoto, R. (1961*b*). On the dominance order in *Polistes* wasps. II. *Scientific Bulletin of the Faculty of of Agriculture of Kyushu University* **19**, 1–17, (in Japanese with English summary).

Moritz, R. F. A. (1986). Two parthenogenetical strategies of laying workers in populations of the honeybee, *Apis mellifera* (Hymenoptera: Apidae). *Entomologia Generalis* **11**, 159–64.

Naumann, M. G. (1970). The nesting behavior of *Protopolybia pumila* in Panama (Hymenoptera: Vespidae). Unpublished Ph.D. thesis, University of Kansas, Lawrence.

Noonan, K. M. (1978). Sex ratio of parental investment in colonies of the social wasp *Polistes fuscatus*. *Science* **199**, 1354–6.

Noonan, K. M. (1981). Individual strategies of inclusive-fitness-maximizing in *Polistes fuscatus* foundresses. In *Natural selection and social behavior: recent research and new theory* (ed. R. D. Alexander and D. W. Tinkle), pp. 18–44. Chiron Press, New York.

Ohgushi, R., Sakagami, S., Yamane S., and Abbas, N. D. (1983). Nest architecture and related notes of stenogastrine wasps in the province of Sumatera Barat, Indonesia (Hymenoptera: Vespidae). *Scientific Report of Kanazawa University* **28**, 27–58.

Oster, G. H. and Wilson E. O. (1978). *Caste and ecology in the social insects*. Princeton University Press, Princeton.

Packer, L. (1986). Multiple foundress associations in a temperate population of *Halictus ligatus* (Hymenoptera: Halictidae). *Canadian Journal of Zoology* **64**, 2325–32.

Pagden, H. T. (1976). A note on colony founding by *Ropilidia* (*Icarielia timida* van der Vecht). *Ned. Akad. Wettensch., Proc. (C) Biol., Medic. and Sci.* **79**, 506–9.

Page, R. E. (1986). Sperm utilization in social insects. *Annual Review of Entomology* **31**, 297–320.

Pamilo, P. and Crozier, R. H. (1982). Measuring genetic relatedness in natural populations: methodology. *Theoretical Population Biology* **21**, 171–93.

Pardi, L. (1942). Richerche sui Polistini. 5. La poliginia initiiale di *Polistes gallicus* (L.). *Bollettino dell'Istituto di Entomologia dell'Universitá di Bologna* **14**, 1–106.

Pardi, L. (1946). Richerche sui Polistini. VII. La 'dominazione' e il ciclo ovarico annuale in *Polistes gallicus* (L.). *Bollettino dell'Istituto di Entomologia dell' Universitá di Bologna* **15**, 25–84.

Pardi, L. (1948). Dominance order in *Polistes* wasps. *Physiological Zoology* **21**, 1–13.

Pardi, L. and Marino Piccioli, M. T. (1970). Studi sulla biologia di *Belonogaster* (Hymenoptera: Vespidae). 2. Differenziamento castale incipiente in *B. griseus* (Fab.). *Monitore Zoologico Italiano (N. S.) Suppl. III.* **11**, 235–65.

Peeters, C. and Crew, R. (1985). Worker reproduction in the ponerine ant *Ophthalmopone berthoudi*: an alternative form of eusocial organization. *Behavioral Ecology and Sociobiology* **18**, 29–37.

Pickering J. (1980). Sex ratio, social behavior and ecology in *Polistes* (Hymenoptera: Vespidae), *Pachysomoides* (Hymenoptera: Ichneumonidae) and *Plasmodium* (Protozoa, Haemosporida). Unpublished Ph.D. thesis, Harvard University, Cambridge, Massachusetts.

Queller, D. C., Strassmann, J. E. and Hughes, C. R. (1988). Genetic relatedness in colonies of tropical wasps with multiple queens. *Science* **242**, 1155–7.

Rau, P. (1933). *The jungle bees and wasps of Barro Colorado Island*. Kirkland, St. Louis Co., Missouri.

Rau, P. (1943). The nesting habits of Mexican social wasps and solitary wasps of the family Vespidae. *Annals of the Entomological Society of America* **36**, 515–36.

Richards, O. W. (1969). The biology of some African social wasps. *Memorie della Società di Entomologia, Italy* **48**, 79–93.

Richards, O. W. (1978*a*). *The social wasps of Americas excluding the Vespidae*. British Museum (Natural History), London.

Richards, O. W. (1978*b*). The Australian social wasps (Hymenoptera: Vespidae). *Australian Journal of Zoology, Suppl. Ser.* No. **61**, 1–132.

Richards, O. W. and Richards, M. J. (1951). Observations on the social wasps of South America (Hymenoptera: Vespidae). *Transactions of the Royal Entomological Society London* **102**, 1–170.

Rissing, S. W. and Pollock, G. B. (1986). Social interaction among pleometrotic queens of *Veromessor pergandei* (Hymenoptera: Formicidae) during colony foundation. *Animal Behaviour* **34**, 226–33.

Rissing, S. W. and Pollock, G. B. (1987). Queen aggregation pleometrotic advantage and brood raiding in the ant *Veromessor pergandei* (Hymenoptera: Formicidae). *Animal Behaviour* **35**, 975–81.

Rissing, S. W. and Pollock, G. B. (1988). Pleometrosis and polygyny in ants. In *Interindividual behavioral variability in social insects* (ed. R. L. Jeanne), pp. 179–222. Westview Press, Boulder, Colorado.

Rissing, S. W., Pollock, G. B., Higgins, M. R., Hagen, R. H. and Smith, D. R. (1989). Foraging specialization without relatedness or dominance among cofounding ant queens. *Nature* **338**, 420–2.

Robinson, G. E. and Page, R. E. (1988). Genetic determination of guarding and undertaking in honeybee colonies. *Nature* **333**, 356–8.

Roisin, Y. (1987). Polygyny in *Nastitermes* species: field data and theoretical approaches. In *From individual to collective behavior in social insects* (ed. J. M. Pasteels and J-L. Deneubourg), pp. 379–404. Birkhäuser Verlag, Basle.

Roisin, Y. and Pasteels, J. M. (1985). Imaginal polymorphism and polygyny in the New-Guinean termite *Nastitermes princeps* (Desneux). *Insectes Sociaux* **32**, 146–57.

Röseler, P-F. (1985). Endocrine basis of dominance and reproduction in polistine wasps. In *Experimental behavioral ecology* (ed. B. Hölldobler and M. Lindauer), pp. 259–72. Gustav Fischer, Stuttgart.

Rosengren, R. and Pamilo, P. (1983). The evolution of polygyny and polydomy in mound-building *Formica* ants. *Acta Entomologica Fennica* **42**, 65–77.

Ross, K. G. and Matthews, R. W. (eds.) (1991). *The social biology of wasps*. Cornell University Press, Ithaca, New York.

Roubaud, E. (1916). Recherches biologiques sur les guêpes solitaires et sociales d'Afrique. *Annales des Sciences Naturelles, Zoologie et Biologie Animale* **10**, 100–60.

Sakagami, S. F. (1970). *A road followed by honeybees.* Sisakusya, Tokyo (in Japanese).

Sakagami, S. F. (1984). *The world of honeybees.* Iwanami, Tokyo, (in Japanese).

Sakagami, S. F. and Maeta, Y. (1985). Multifemale nests and rudimentary castes in the normally solitary bee *Cerattis japonica* (Hymenoptera: Xylocopidae). *Journal of the Kansas Entomological Society* **57**, 635–54.

Schwarz, M. P. (1986). Persistent multi-female nests in an Australian allodapine bee, *Exoneura bicolor* (Hymenoptera: Anthophoridae). *Insectes Sociaux* **33**, 258–77.

Seger, J. (1983). Partial bivoltinism may cause alternating sex-ratio biases that favour eusociality. *Nature* **201**, 59–62.

Sekiguchi, K. and Sakagami, S. F. (1966). Structure of foraging population and related problems in the honeybees, with consideration on the division of labour in bee colonies. *Report No. 69, Hokkaido National Agricultural Experiment Station*, pp. 1–65.

Sekijima, M., Sugiura, M., Matsuura, M. (1980). Nesting habits and brood development of *Parapolybia indica* Saussure (Hymenoptera: Vespidae). *Bulletin of the Faculty of Agriculture of Mie University* **61**, 11–23.

Siew, Y. S. and Sudderuddin, K. I. (1982). Malaysian hornets—some interesting facts. *Nature Malaysiana* **7**, 18–21.

Skaife, S. H. (1953). Subsocial bees of the genus *Allodape* Lep. and Serv. *Journal of the Entomological Society of South Africa* **34**, 251–71.

Snelling, R. R. (1981). Systematics of social Hymenoptera. In *Social insects* Vol. 2 (ed. H. R. Hermann), pp. 369–453. Academic Press, New York.

Spradbery, J. P. (1973). The European social wasp, *Paravespula germanica* (F.) (Hymenoptera: Vespidae) in Tasmania, Australia. *IUSSI Proceedings of the VII International Congress*, pp. 375–80.

Spradbery, J. P. (1975). The biology of *Stenogaster concinua* van der Vecht with comments on the phylogeny of Stenogastrinae (Hymenoptera: Vespidae). *Journal of the Australian Entomological Society* **14**, 309–18.

Spradbery, J. P. (1986). Polygyny in the vespinae with special reference to the hornet *Vespa affinis picea* Buysson (Hymenoptera: Vespidae) in New Guinea. *Monitore Zoologico Italiano* (*N.S.*) **20**, 101–18.

Spradbery, J. P. (1991). Evolution of queen number and queen control. In *Social biology of wasps* (ed. K. G. Ross and R. W. Matthews), pp. 336–88. Cornell University Press, Ithaca, New York.

Starr, C. K. (1976). Nest reutilization of *Polistes metricus* (Hymenoptera: Vespidae) and possible limitation of multiple foundress associations by parasitoids. *Journal of the Kansas Entomological Society* **49**, 142–4.

Strassmann, J. E. (1981*a*). Evolutionary implications of early male and satellite nest production in *Polistes exclamans* colony cycles. *Behavioural Ecology and Sociobiology* **8**, 55–64.

Strassmann, J. E. (1981*b*). Wasp reproduction and kin selection: Reproductive competition and dominance hierarchies among *Polistes annularis* foundresses. *Florida Entomologist* **64**, 74–88.

Strassmann, J. E. (1983). Nest fidelity and group size among foundresses of *Polistes annularis* (Hymenoptera: Vespidae). *Journal of the Kansas Entomological Society* **56**, 621–34.

Strassmann, J. E. (1989*a*). Group colony foundation in *Polistes annularis* (Hymenoptera: Vespidae). *Psyche* **96**, 223–36.

Strassmann, J. E. (1989*b*). Altruism and relatedness at colony foundation in social insects. *Trends in Ecology and Evolution* **4**, 371–4.

Strassmann, J. E., Queller, D. C., and Hughes, C. R. (1988). Predation and the evolution of sociality in the paper wasp *Polistes bellicosus*. *Ecology* **69**, 1497–505.

Strassmann, J. E., Hughes, C. R., Queller, D. C., Turillazzi, S., Cervo, R., Davis, S. K., and Goodnight K. F. (1989). Genetic relatedness in primitively eusocial wasps. *Nature* **342**, 268–70.

Sturtevant, A. H. (1938). Essays on evolution. II. On the effects of selection on social insects. *Quarterly Review of Biology* **13**, 74–6.

Suzuki, T. (1985). Mating and laying of female-producing eggs by orphaned workers of a paper wasp, *Polistes snelleni* (Hymenoptera: Vespidae). *Annals of the Entomological Society of America* **78**, 736–9.

Suzuki, T. (1987). Egg-producers in the colonies of a polistine wasp, *Polistes snelleni* (Hymenoptera: Vespidae), in central Japan. *Ecological Research* **2**, 185–9.

Thorne, B. L. (1982*a*). Polygyny in termites: multiple primary queens in colonies of *Nastutitermes corniger* (Motschulsky) (Isoptera: Termitidae). *Insectes Sociaux* **29**, 102–117.

Thorne, B. L. (1982*b*). Multiple primary queens in termites: phyletic distribution, ecological context and a comparison to polygyny in Hymenoptera. In *The biology of social insects* (ed. M. D. Breed, C. D. Michener, and H. E. Evans), pp. 206–11. Westview Press, Boulder, Colorado.

Thorne, B. L. (1984). Polygyny in the Neotropical termite *Nasutitermes corniger*: life history consequences of queen mutualism. *Behavioural Ecology and Sociobiology* **14**, 117–36.

Trivers, R. L. (1974). Parent-offspring conflict. *American Zoologist* **14**, 249–64.

Tschinkel, W. R. and Howard D. F. (1983). Colony founding by pleometrosis in the fire ant. *Solenopsis invicta*. *Behavioural Ecology and Sociobiology* **12**, 103–13.

Tsuji, K. (1988). Obligate parthenogenesis and reproductive division of labor in the Japanese queenless ant *Pristomyrmex pungens*: comparison of intranidal and extranidal workers. *Behavioural Ecology and Sociobiology* **23**, 247–55.

Tsuji, K. (1990). Reproductive division of labour related to age in the Japanese queenless ant, *Pristomyrmex pungens*. *Animal Behaviour* **39**, 843–9.

Tsuneki, K. (1965). The biology of East-Asiatic *Cerceris* (Hymenoptera: Sphecidae) with special reference to the peculiar social relationships and return to the nest of *Cerceris hortivaga* Kohl. *Etizenia* **9**, 1–46.

Turillazzi, S. (1986). Colony composition and social behaviour of *Parischnogaster alternata* Sakagami (Hymenoptera: Stenogastrinae). *Monitore Zoologico Italiano* (*N.S.*) **20**, 333–47.

Turillazzi, S. (1987). Social biology of the Stenogastrinae: Temporary dynamic reproductive strategy in the wet tropics. In *Chemistry and biology of social insects* (ed. J. Eder and H. Rembold), pp. 381–2. Verlag J. Peperny, München.

Turillazzi, S., Marino Piccioli, M. T., Hervatin, L. and Pardi, L. (1982). Reproductive capacity of single foundress and associated foundress females of *Polistes gallicus* (L.) (Hymenoptera: Vespidae). *Monitore Zoologico Italiano* (*N.S.*) **16**, 75–88.

Turillazzi, S. and Pardi, L. (1982). Social behavior of *Parischnogaster nigricans serrei* (Hymenoptera: Vespidae) in Java. *Annals of the Entomological Society of America* **3**, 657–64.

Turillazzi, S. and Turillazzi, M. (1985). Notes on the social behaviour of *Ropalidia fasciata* (F.) in West Java (Hymenoptera: Vespidae). *Monitore Zoologico Italiano* (*N.S.*) **19**, 219–30.

Vehrencamp, S. L. (1983). A model for the evolution of despotic versus egalitarian societies. *Animal Behaviour* **31**, 667–82.

Vehrencamp, S. L. (1984). Exploitation in co-operative societies: models of fitness biassing in co-operative breeders. In *Producers and scroungers: strategies of exploitation and parasitism* (ed. C. J. Bernard), pp. 229–66. Croom Helm, London.

Waloff, N. (1957). The effect of the number of queens of the ant *Lasius flavus* (Fab.) (Hym., Formicidae) on their survival and on the rate of development of the first brood. *Insects Sociaux* **4**, 391–408.

Ward, P. S. (1983a). Genetic relatedness and colony organization in a species complex of ponerine ants. I. Phenotypic and genotypic composition of colonies. *Behavioural Ecology and Sociobiology* **12**, 285–99.

Ward, P. S. (1983b). Genetic relatedness and colony organization in a species complex of ponerine ants. II. Patterns of sex ratio investment. *Behavioural Ecology and Sociobiology* **12**, 301–307.

Wcislo, W. T., West-Eberhard, M. J. and Eberhard, W. G. (1988). Natural history and behavior of a primitively social wasp, *Auplopus semialatus*, and its parasite, *Irenangelus eberhardi* (Hymenoptera: Pompilidae). *Journal of Insect Behavior* **1**, 247–67.

Wenzel, J. W. (1987). *Ropalidia formosa*, a nearly solitary paper wasp from Madagascar (Hymenoptera: Vespidae). *Journal of the Kansas Entomological Society* **60**, 549–56.

West-Eberhard, M. J. (1969). The social biology of polistine wasps. *University of Michigan Museum of Zoology Miscellaneous Publications* **140**, 1–101.

West-Eberhard, M. J. (1973). Monogyny in 'polygynous' social wasps. *Proceedings of the VII Congress IUSSI (London)*, pp. 396–403.

West-Eberhard, M. J. (1978a). Temporary queens in *Metapolybia* wasps: non-reproductive helpers without altruism? *Science* **200**, 441–3.

West-Eberhard, M. J. (1978b). Polygyny and the evolution of social behavior in wasps. *Journal of the Kansas Entomological Society* **51**, 832–56.

West-Eberhard, M. J. (1982a). The nature and evolution of swarming in tropical social wasps (Vespidae, Polistinae, Polybiini). In *Social insects in the tropics* (ed. P. Jaisson), Vol. 1, pp. 97–128. University of Paris XIII Press, Paris.

West-Eberhard, M. J. (1982b). Communication in social wasps: predicted and observed patterns, with a note on the significance of behavioral and ontogenetic flexibility for theories of worker 'altruism'. In *La communication chez les sociétés d'insectes* (ed. A. de Haro), pp. 13–35. Ballaterra, Barcelona.

West-Eberhard, M. J. (1986). Dominance relations in *Polistes canadensis* (L.), a tropical wasp. *Monitore Zoologico Italiano (N.S.)* **20**, 263–81.

West-Eberhard, M. J. (1987). Flexible strategy and social evolution. In *Animal societies: theories and facts* (ed. Y. Itô, J. L. Brown and J. Kikkawa), pp. 35–51. Japanese Science Society Press, Tokyo.

Wheeler, W. M. (1923). *Social life among the insects.* Harcourt Brace, New York.

Wilson, E. O. (1966). Behaviour of social insects. *Symposium of the Royal Entomological Society of London* **3**, 81–96.

Wilson, E. O. (1971). *The insect societies.* Belknap Press of Harvard University Press, Cambridge, Massachusetts.

Wilson, E. O. (1975). *Sociobiology: the new synthesis.* Belknap Press of Harvard University Press, Cambridge, Massachusetts.

Woyke, J. (1975). Natural and instrumental insemination of *Apis cerana indica* in India. *Journal of Apicultural Research* **14**, 153–69.

Wright, S. (1945). Tempo and mode in evolution: a critical review. *Ecology* **26**, 415–19.

Yamamura, N. (1986). *Mathematical models of reproductive strategies.* Tôkai University Press, Tokyo, (in Japanese).

Yamane, S. (1980). Social biology of the *Parapolybia* wasps in Taiwan. D. Sci. dissertation, pp. 213, Hokkaido University, Sapporo.

Yamane, S. (1985). Social relations among females on pre- and post-emergence colonies of a subtropical paper wasp, *Parapolybia varia* (Hymenoptera: Vespidae). *Journal of Ethology* **3**, 27–38.

Yamane, S. (1986). The colony cycle of the Sumatran paper wasp, *Ropalidia (Icariola) variegata jacobsoni* (Buisson), with reference to the possible occurrence of serial polygyny (Hymenoptera: Vespidae). *Monitore Zoologico Italiano* (*N.S.*) **20**, 135–61.

Yamane, S., Kojima, J. and Yamane, Sk. (1983*a*). Queen worker size dimorphism in an Oriental polistine wasp, *Ropalidia montana* Carl (Hymenoptera: Vespidae). *Insectes Sociaux* **30**, 416–22.

Yamane, S., Sakagami, S. F. and Ohgushi, R. (1983*b*). Multiple behavioral options in a primitively social wasp, *Parischnogaster mellyi*. *Insectes Sociaux* **30**, 412–15.

Yamane, S., Itô, Y. and Spradbery, J. P. (1991). Comb cutting in *Ropalidia plebeiana*: a new process of colony fission in social wasps (Hymenoptera: Vespidae). *Insectes Sociaux* **38**, 105–10.

Yamane, Sk. (1973). Discovery of a pleometrotic association in *Polistes chinensis antennalis* Pérez (Hymenoptera: Vespidae). *Life Study* **17**, 3–4.

Yoshikawa, K. (1957). A brief note on the temporary polygyny in *Polistes fadwigae* Dalla Torre, the first discovery in Japan. *Mushi* **30**, 37–9.

Yoshikawa, K. (1962). Introductory studies on the life economy of polistine wasps. VII. Comparative consideration and phylogeny. *Journal of Biology, Osaka City University* **13**, 45–64.

Yoshikawa, K. (1963). Introductory studies on the life economy of polistine wasps. II. Superindividual stage. 3. Dominance order and territory. *Journal of Biology, Osaka City University* **14**, 55–61.

Yoshikawa, K. Ohgushi, R. and Sakagami, S. F. (1969). Preliminary report on entomology of the Osaka City University 5th Scientific Expedition to Southeast Asia 1966. *Nature and Life in Southeast Asia* **6**, 153–82.

Author index

Index of scientific names

Subject index

ACS-4981 3/20/95